THE SCIENCE
OF ALiENS

Previous Works by Clifford Pickover

The Alien IQ Test

Black Holes, A Traveler's Guide

Chaos in Wonderland: Visual Adventures in a Fractal World

Chaos and Fractals: A Computer-Graphical Journey

Computers, Pattern, Chaos, and Beauty

Computers and the Imagination

Future Health: Computers and Medicine in the 21st Century

Fractal Horizons: The Future Use of Fractals

Frontiers of Scientific Visualization (with Stu Tewksbury)

Keys to Infinity

The Loom of God

Mazes for the Mind: Computers and the Unexpected

Mit den Augen des Computers

The Pattern Book: Fractals, Art, and Nature

Spider Legs (with Piers Anthony)

Spiral Symmetry (with Istvan Hargittai)

Strange Brains and Genius

Time, A Traveler's Guide

*Visions of the Future: Art, Technology, and Computing
in the 21st Century*

Visualizing Biological Information

THE SCIENCE OF ALIENS

CLIFFORD A. PICKOVER

BASIC
BOOKS

A Member of the Perseus Books Group

Copyright © 1998 by Clifford A. Pickover.

Published by Basic Books,
A Member of the Perseus Books Group

All rights reserved. Printed in the United States of America. No part of this book
may be reproduced in any manner whatsoever without written permission except in
the case of brief quotations embodied in critical articles and reviews. For information, address Basic Books, 10 East 53rd Street, New York, NY 10022–5299.

FIRST EDITION

Designed by Rachel Hegarty

A CIP catalog record for this book is available from the Library of Congress.
ISBN 0-465-07314-X

98 99 00 01/❖RDM   10   9   8   7   6   5   4   3   2   1

This book is dedicated to the cheela,
hyperintelligent slugs dwelling
on a distant neutron star.

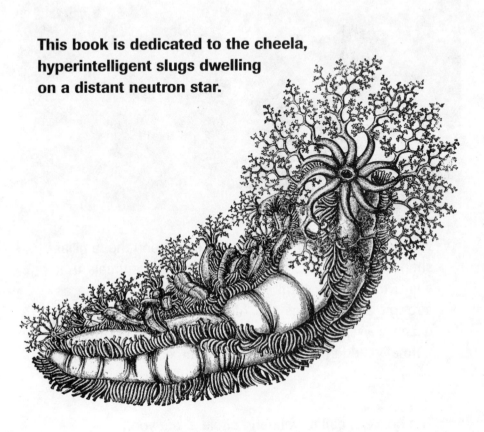

Nowhere in space will we rest our eyes upon the familiar shapes of trees and plants, or any of the animals that share our world. Whatsoever life we meet will be as strange and alien as the nightmare creatures of the ocean abyss, or of the insect empire whose horrors are normally hidden from us by their microscopic scale.

—Arthur C. Clarke, 1962

The heavens call to you, and circle about you,
displaying to you their eternal splendors,
and your eye gazes only to earth.

—Dante, 1300

How would it be if we discovered that aliens
only stopped by earth to let their kids take a leak?

—Jay Leno, 1997

CONtENTS

ACkNOWLEDGMENTS

We are in the position of a little child entering a huge library whose walls are covered to the ceiling with books in many different tongues. . . . The child does not understand the languages in which they are written. He notes a definite plan in the arrangement of books, a mysterious order which he does not comprehend, but only dimly suspects.

—Albert Einstein

Far-away planets have their plants and animals, and their rational creatures too, and those as great admirers, and as diligent observers of the heavens as ourselves.

—Christiaan Nuygens, seventeenth-century Dutch physicist and astronomer

Cosmology is where science and religion meet.

—George Smoot, 1992

I thank Brad Marshall for the full-page alien illustration on p. xiv in the Preface and Eduardo Abel Gimenez for the alien squid-men on the opening quotation page. The Persian calligraphy in the Introduction is written by Jalil A. Taghizadeh. Figure 10.1 is courtesy of Stelarc, and the photographers include P. Fernuik and S. Hunter. Several figures of Earthly animals come from the Dover Pictorial Archive.

I thank Dr. H. Paul Shuch, the Executive Director of the SETI League, Dr. Jack Cohen of the University of Warwick, Craig Becker, Dina Roumiantseva, Martie Saxenmeyer, Max Rible, Johan Forsberg, Susan Rabiner, Jim McLean, Greg Kishi, Mike Hocker, Dave Glass, and Dan Winarski for their useful comments and encouragement. I thank Dave Roberts of the Department of Zoology, the Natural History Museum, London, for information on extremophiles.

Readers are encouraged to consult Wayne Barlowe's *Barlowe's Guide to Extraterrestrials* for excellent color drawings of famous science fiction aliens, and Doris and David Jonas's *Other Senses, Other Worlds* for more information on alien senses. Walter Sullivan's *We Are Not Alone* and Frank Drake's and Dava Sobel's *Is Anyone Out There?* provide fascinating information on SETI, the Search for Extraterrestrial Intelligence.

PREfACE

To my utter astonishment I saw an airship descending over my cow lot. It was occupied by six of the strangest beings I ever saw. They were jabbering together, but we could not understand a word they said.

—Congressman Alexander Hamilton, 1897

If we wish to understand the nature of the Universe we have an inner hidden advantage: we are ourselves little portions of the universe and so carry the answer within us.

—Jacques Boivin, *The Single Heart Field Theory*

Astronomy compels the soul to look upwards and lead us from this world to another.

—Plato (427–347 B.C.), *The Republic,* Book VII

Are We Alone?

I first became obsessed with the notion of alien life-forms as a child watching black-and-white episodes of the 1960s TV series *The Outer Limits*. You can't imagine how profoundly affected I was by the blurring of fact and fiction. The strange array of aliens, from the antlike Zanti Misfits to the noble Galaxy Being, made the unbelievable seem a frighteningly real possibility.

My interest was further stimulated in the late 1960s by the TV series *Lost in Space,* which dealt with the travels of a human family exploring strange planets. Their mission, set in the not-too-distant future, was to begin colonizing a planet near the star Alpha Centauri. Unfortunately, their craft went off course, and they lost all contact with Earth.

The most memorable *Lost in Space* episode dealt with an intergalactic zoo. When the animals accidentally escape, an odd assortment of

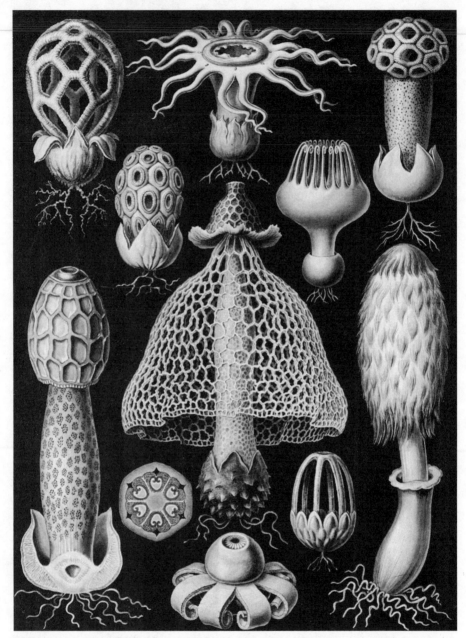

Various species of the class Basidiomycetes.

two-legged, hairy, bug-eyed monsters come ambling, running, leaping, and shuffling from the zoo enclosure. It hardly occurred to me to ask myself what would it *really* be like to visit an intergalactic zoo. Would the creatures look like the ones on the TV show?

What if *you* could visit an intergalactic zoo filled with intelligent life-forms? Would the aliens have heads, arms, and legs, or even be vaguely humanoid? The challenging task of imagining aliens from other worlds is useful for any species that dreams of understanding its place in the universe.

Are humans alone in the universe? This question is one of the oldest questions posed by philosophers and scientists, and it has profound implications for our worldview. For the first time in history, questions about extraterrestrial life have left the realm of theology and science fiction and entered the realm of experimental science. Recent advances in biochemistry and molecular biology suggest that life—even life on Earth—can exist in incredibly diverse and bizarre environments. Recent discoveries of life living miles under the earth in utter darkness, or in ice, or even in boiling water, tell us that that which is *possible* in nature tends to become *realized*. My personal view is that almost everything happens in our universe that is not forbidden by the laws of physics and chemistry. Life on Earth can thrive in unimaginably harsh conditions, even in acid or within solid rock. On the ocean floor, bacteria thrive in scalding, mineral-laden hot springs. *If microbes thrive in such miserable conditions on Earth, where else beyond Earth might similar life-forms exist?*

As our understanding of the origin of life and the chemical composition of our Solar System's planets and moons increases, we learn that life on other worlds is quite likely, and we will someday discover it in myriad forms. Compelling evidence of a subsurface ocean on Jupiter's moon Europa has also fueled speculation that life can exist on other worlds. Moreover, the wonderful and extremely weird creatures that inhabit our own world suggest that God loves the *bizarre*. When I gaze upon crazy-looking crustaceans; squishy-tentacled jelly-fish; grotesque, hermaphroditic worms; and Basidiomycetes fungi more alien than the wildest dreams of science fiction writers,[1] I know that God has a sense of humor, and we will see this reflected in other forms in the universe.

The Alien Smorgasbord

I've attempted to make *The Science of Aliens* a strange journey that unlocks the doors of your imagination with thought-provoking mysteries, puzzles, and problems on topics ranging from alien senses, parallel universes, and alien sex to alien abduction. A resource for science fiction aficionados, a playground for philosophers, an adventure and education for science students, each chapter is a world of paradox and mystery.

There are many excellent books on the possibility of extraterrestrial life, and these are listed in the section "For Further Reading" at the end of this book. So, why another book on alien life? I have found that previous books on the subject have a particular shortcoming. They don't focus wholeheartedly on the physical appearance, philosophy, and evolution of aliens while also discussing leading-edge biological research on Earthly life.

I hope that my army of illustrators will also stimulate your imagination in ways that mere words cannot. Imagery is at the heart of much of the work described in this book. To better understand and contemplate alien life, we need to use our eyes. Artists from different backgrounds produce visual representations from myriad perspectives. For many of you, seeing hypothetical aliens will clarify concepts in ways that words alone cannot.

Why contemplate the appearance of aliens? Scientists and artists feel the excitement of the creative process when they leave the bonds of the known to venture far into unexplored territory lying beyond the prison of the obvious. When we envision the physical structures of alien bodies and their hypothetical cultures, we are at the same time holding a mirror to ourselves, revealing our own prejudices and preconceived notions. Aliens appeal to young minds, and I know of no better way to stimulate students than to muse about the science of aliens. Creative minds love roaming freely through alien biology, psychology, and astronomy.

As with all my previous books, you are encouraged to pick and choose from the smorgasbord of topics. Feel free to skip chapters in favor of those topics of most interest to you. A few morsels of information are repeated so that each chapter contains sufficient background information. In fact, there is nothing to stop you from reading this book from back to front. Many of the chapters are brief, to give you just the tasty flavor of a topic. Those of you interested in pursuing specific topics can find additional information in the referenced publications. In order to encourage your involvement, I have loaded the book with numerous what-if questions for further thought. Spread the spirit of this book by posing these questions to your students, to your buddies at the next stockholders' meeting, to your family the next time you plunk down on the couch to watch *Star Wars*.

INTRODUCTION

UFO mythology is similar to the message of the classical religions where God sends his Angels as emissaries who offer salvation to those who accept the faith and obey his Prophets. Today, the chariots of the gods are UFOs. What we are witnessing in the past half century is the spawning of a New Age religion.

—Paul Kurtz

Know thou that every fixed star hath its own planets, and every planet its own creatures, whose number no man can compute.

—Baha'u'llah' (1817–1892)

● ● ●

FBI Headquarters, Medical Wing A—9:00 P.M.

You are in a steam-sterilized room assisting FBI agents Fox Mulder and Dr. Dana Scully of *The X-Files*. On a gleaming stainless-steel autopsy table is something you came upon only hours before in the woods by your home—a four-foot-tall, pea-green alien. Although apparently dead, its eyes still sparkle. From beneath its thin, moist mouth, a hundred throat appendages quiver aperiodically—presumably some residual activity in local nerves.

Dana Scully dons a white mask. "Let's open the head," she says. "I want to have a look at the brain."

Fox Mulder wrinkles his nose as the alien's gray-green eyelids ooze a tiny amount of liquid that smells like wasabi. "I'm not sure I want to watch," he says to Scully.

"You'll be fine. We should operate before rigor mortis sets in."

Mulder looks at her with concern in his eyes. "Okay, what next?"

"Get the bone drill."

You assist Mulder by helping him find a small battery-powered drill from a nearby cabinet. Scully dons latex gloves, removes a scalpel from an instrument tray, and gently slides it across the forehead of the alien. Blood wells up in a sudden chartreuse thread as she continues the incision along the alien's forehead.

"Here goes," Scully says as she fires up the bone drill with a *wrrr*.

The Huntington electric bone drill begins to hum. She holds the tip of the drill to a point on the alien's skull about one inch above its eyes. "Damn," she says, "this skull is thick." She presses down harder, and the drill starts to make a high-pitched grinding noise. Musty smoke, filled with tiny bone particles, rises into the cool air.

After making several holes, Scully replaces the drill burr with a circular cutting disk. For a second, the saw begins to jump away as it cuts into the bone, but Scully's grip holds firm. In a few minutes she is able to remove a neatly sectioned portion of the alien's cranial cap.

Scully lifts the bone cup gently, and it makes a popping sound like a cork being blown from a champagne bottle.

You look closer. "Looks vaguely human," you say, gazing at the shiny wrinkled jelly of the alien brain.

Scully nods. "Yes, the brain's structure is symmetrical."

The small interlocking ridges on the surface of the cerebral cortex remind you of the deep fissures and protrusions of a delicious mango— but you keep that observation to yourself.

"Look," Scully says, "there're two huge optic nerve tracts carrying fibers from the eyes. And here's the equivalent of the pineal body. Descartes thought it was the seat of the human soul."

Mulder comes closer. "You mean Alfonso Descartes from food service?"

"Not likely. Talking philosophy here."

Scully pauses and takes a deep breath. "Let's take a look at the ventricles," she says. "If this creature is anything like us, it should have four ventricles forming a network of interconnecting cavities. They should be filled with cerebral spinal fluid to cushion the brain."

Scully reaches in deeper and looks at you. "Could you help hold apart the lobes as I cut?"

"Yes," you agree apprehensively. You've never touched an alien.

She slices into the brain, revealing the ventricle chambers.

Mulder takes a peek. "Scully, what do you think of that? They're not filled with fluid. Just empty."

"Look!" she screams.

You pry apart the cavity further. Inside one of the brain ventricles is a little hairy creature about the size of a dime. Occasionally its hair parts, and you see it has eyes and a mouth. A second creature in another ventricle wanders back and forth along a complex of fibers resembling cables. Both creatures stop whatever they are doing and gaze up at you from their home in the brain.

"Get the hell out of here," one of them says with a strange accent. The other whispers, "Evett mar valaki ebbol az etelbol amit nekem kinalsz vagyen vagyok az elso?" Both of the little creatures stare at you. This is one of those rare occasions during which you nearly faint.

"My God!" Mulder says. "Get a jar to capture them!"

You feel a warm shiver along your neck as you look into the creatures' coal-black eyes. You feel a fire, an ambiguity, a creeping despair. The tiny creatures never move, never even blink. Their eyes are dark, their smiles relentless and practiced. Time seems to stop. For a moment, you feel quite hot. But when you shake your head, the heat is gone. Just your imagination. But the creatures remain. Cruel. Nightmarish. You feel as though you are caught in a subterranean dungeon and the door suddenly closed.

Scully drops the brain as the two creatures scurry back inside the ventricles. As the brain hits the tile floor, you hear a pop as the cerebellum implodes.

Regaining your composure, you stoop down, reopen the brain, and ask the tiny creatures the obvious, "What are you?" You hold the brain as far away as you can. You don't want them to jump out at your face.

"I am called Ka," one says.

"I am called Da," says the other.

You look at Mulder and Scully, unsure of how to proceed. Scully whispers to Mulder, "Hand me a jar. Quickly!"

"What is your purpose?" you ask the little creatures.

The creatures look at each other for a minute as if to choose among interpretations. . . .

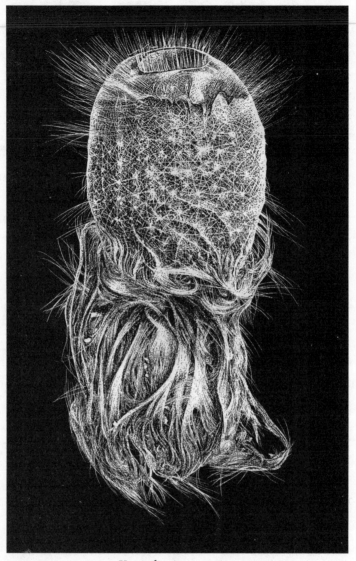

Ka, a brain parasite.

Da motions to the pea-green alien. "All Prohaptors have a Ka and Da in their brains to help perform maintenance activities. We help distribute oxygen, repair damage, help permanently implant memories, and regulate emotions and various hormones."

You turn to Scully. "Do you think we should take them as specimens?" you say.

"'Take them as specimens?'" Da echoes you and then turns to Ka. "What in God's name is he talking about?" Da says to Ka.

Scully nods. "We'll learn a lot by studying them," she says. "But I'd bet they'll soon die now that their host is dead."

You bring out some forceps and reach for them.

"Hey!" Ka says. "What do you think you're doing?"

Both Ka and Da retreat into the depths of the dead brain.

"Be careful," Scully says to you. They might bite."

You hear a voice coming from somewhere in the wet organ. "Yeah," it says, "that's right, we might bite."

You start slicing pieces off of the brain. Soon there is nowhere they can run. Nowhere they can hide.

You pick up Ka and Da with forceps as they scream at the three of you.

●   ●   ●

Just how unrealistic is this admittedly zany scenario? Would aliens even have brains? Would the aliens look like little green humans, or be more exotic, such as the creature just described?

Over its several seasons, *The X-Files* TV show has had plenty of little green man–type aliens: At least eight classic extraterrestrials (which, by the way, lacked nipples and belly buttons) have been designed for the Fox series by Toby Lindala, head of makeup effects. Pop culture abounds. Just how plausible are aliens that resemble the big-eyed humanoid aliens found on the covers of such books as Whitley Strieber's *Communion*? During this hypothetical autopsy scene, a creature converses with you in your native language. What are the chances that we could learn to communicate at all with an extraterrestrial? What *would* aliens look like? Would they have internal organs like our own? We'll encounter all these and other questions as we open doors. . . .

Opening the Doors

If space-faring aliens are circling some of the stars closest to Earth, by now they should have detected our radio and TV emissions. However, at present there seems to be no clear physical evidence that aliens are visiting Earth, although we certainly have a strong emotional yearning to believe in extraterrestrial life. Most of us even have some preconceived notion as to how an extraterrestrial life-form behaves and what it looks like. We also arbitrarily equate advanced technology with advanced social development. Perhaps one of the most important lessons of this book is that what we think or imagine about aliens not only reflects our own fears and wishes, but alters them.

Humanity's interest in the extraterrestrial is at least as old as the late Stone Age, when Europeans engraved lunar markings on primitive calendars made of ivory. Gods from nearly every culture have been identified with planets. Wondering about the stars has been part of the intellectual life of every civilization on Earth.

As a scientist and author, I'm driven by curiosity. I want to know what may be cohabiting our vast universe with us. In this book I speculate on more visceral issues than most authors: I want you to know what aliens might look like, what their sex acts might be like, how they may think and act. I also briefly discuss SETI, the search for extraterrestrial intelligence, which uses antennas aimed into the cosmos, searching for radio signals from other worlds. I would love to have the fascinating job of working for SETI because these projects combine aspects of computer technology, radio astronomy, communication, chemistry, and biology, in the boldest adventure in human history. I agree with the ancient Persian proverb "The seeker is a finder," which suggests we must always search in order to understand our place in our universe. The search is important even if it is unsuccessful. Usually when we look for one thing and are unsuccessful, we often discover something else equally exciting. For example, as radio astronomers look into the cosmos, they have made startling, unexpected discoveries ranging from pulsars and the residual glow of the Big Bang to other signals that we can not yet explain. SETI opens the doors to entirely new realms of human thought and adventure. The doors are open—let's wait and watch for what walks in.

Detection

If aliens are out there, Earthlings have only recently become detectable to them with our introduction of radio and TV in the latter part of this century. Our TV shows are leaking into space as electromagnetic signals that can be detected at enormous distances by receiving devices not much larger than our own radio telescopes.[1] Whether we like it or not, *Melrose Place* is heading to Alpha Centauri,[2] and *Baywatch*, the most popular TV show on our planet, is shooting out to the constellation Orion. What impressions would these shows make on alien minds? It is a sobering thought that one of the early signs of terrestrial intelligence might come from the mouth of Bart Simpson.

Similarly, could the first signal we receive from the stars be the equivalent of *The Three Stooges,* with bug-eyed aliens smashing each other with green-goo pies? What if our first message from the stars were alien pornography that inadvertently leaked out into space? SETI funding would be even more difficult if the Reverend Jerry Falwell and other conservatives found that our first broadcast extraterrestrial message was of a hard-core *Playboy*—and the first images received were of aliens plunging their elephantine proboscises into the paroxysmal esophagus of some nubile, alien marsupial.

As hard as it may be to stomach, our *entertainment* will be our earliest transmissions to the stars. If we ever receive inadvertent transmissions from the stars, it will be *their* entertainment. Imagine this. The entire earth sits breathlessly for the first extraterrestrial images to appear on CNN. One of the preppy-and-perfect news anchors appears on our TVs for instant live coverage. And then, beamed to every home, are the alien equivalents of Pamela Sue Anderson in a revealing bathing suit, Beavis and Butt-head mouthing inanities and expletives, and an MTV heavy-metal band consisting of screaming squids.

This is not such a crazy scenario. In fact, satellite studies show that the Super Bowl football action, which is broadcast from more transmitters than any other signal in the world, might be the most easily detected message from Earth. The first signal from an alien world could be the alien equivalent of a football game.

━━━━━━━━━━━━━━━━━━━━━━━━━━━━━━ ● ● ●━━━

Lesson one: We had better not assess an entire culture solely on the basis of its entertainment.

━━━━━ ● ● ●━━━━━━━━━━━━━━━━━━━━━━━━━━━━━

● ● ●

Lesson two: You can learn a lot about a culture from its entertainment.

● ● ●

Cultures Beyond Earth

Even more unnerving than judging a culture from its entertainment is judging a culture from the habits of subgroups. From our perspective, aliens would certainly have strange ways—but so do our fellow humans. As just one example, consider how aliens might react if they visited Earth and studied the Hiji of Nigeria and Cameroon. According to *Strangest Human Sex: Ceremonies and Customs*, the Hiji never bury a dead community member until they skin the body. First, they sit up the corpse for two days on a platform, with its hands in bowls of peanuts and food grains so the crops will stay fertile. Next a blacksmith visits and, with his fingers, yanks off the body's skin, which he throws into a pot and buries. Then, "the skinless corpse is washed with red juice, smeared with goat's fat, dressed, and carried to a burial site."[3]

● ● ●

Lesson three: Judge not an alien culture from studies of one group.

Lesson four: Our own ways may be as alien to us as an alien's ways.

● ● ●

We would probably be nervous if aliens' first contact were with the Hiji. On the other hand, whom would you like them to first meet? If you could beam a message into space for aliens to hear, what would you choose? If you could send a human emissary whom would you send? Someone like Mother Teresa or Albert Einstein? Jesus or Bill Gates?

Back in the early 1970s, Timothy Leary, famous for his experiments with hallucinogens at Harvard in the early 1960s, also wanted to send emissaries to the stars. In particular, he wanted to finance the construction of a starship to preserve humanity in the event the Earth was destroyed by nuclear annihilation. The ship was to be sufficiently large to hold the 300 most important people in the world, whom he would select, using his own criteria. His starship would be a modern-day Noah's Ark that would establish a new civilization on an Earthlike planet of some nearby star. Alas, when the scientists Carl Sagan and Frank Drake explained to Leary that such an ark would take centuries to reach a suitable destination, Leary had to abandon the project.

Alien Awareness

Aliens will be aware of an entirely different universe than we are, because of the difference between our brains and senses and theirs. We can hardly imagine a gorilla understanding the significance of prime numbers, yet the gorilla's genetic makeup differs from ours by only a few percentage points. These minuscule genetic differences in turn produce differences in our brains. Additional alterations of our brains would admit a variety of profound concepts to which we are now totally closed. It is quite possible that the thoughts of aliens could not be understood because our own brains affect our ability to contemplate alien philosophies. What new aspects of an alien's reality could we absorb with extra cerebrum tissue? Philosophers of the past have admitted that the human mind is unable to find answers to some of the most important questions, but these same philosophers rarely thought that our lack of knowledge was due to an organic deficiency shielding our psyches from higher knowledge.

Even though we would share much of our mathematics with intelligent aliens, there are certainly areas of alien mathematics that would be difficult to understand. If the yucca moth, with only a few ganglia for its brain, can recognize the geometry of the yucca flower from birth, how much of our mathematical capacity is hardwired into the convolutions of our cortex? Obviously, an understanding of specific higher mathematics is not inborn, because acquired knowledge is not inherited, but our mathematical capacity *is* a function of our brain. There is an organic limit to our mathematical depth. There is an organic limit to our ability to understand alien "truths."

How can we predict what kinds of alien life-forms we may someday encounter in the course of interplanetary exploration? One way to imagine this is to consider the forces that produced the diversity of senses and intelligences right here on Earth. In a real way, there are already alien worlds right here among us. Every Earthly creature perceives the world in an "alien" way. Dogs. Bees. Bats. Cats. They experience the world with different kinds of senses. They can smell what we cannot; they can see what we cannot; they can hear what we cannot. If the organisms of Earth were somehow able to describe their world to you, it would probably not be recognizable to you. It would seem like the wildest world from any science fiction story. Moreover, if you were able to describe the world to another species, they would "see" no resemblance to their own. Our sense of reality would be different; our way of thinking would be different; and even the practical technology we would produce would be different. We do not have to contemplate aliens or science fiction to imagine alienlike senses and bodies. The animal world of Earth is so diverse and full of different senses that creatures are already walking among us possessing "alien" awarenesses beyond our understanding.

I sometimes ask my colleagues to imagine what our world would be like if all rodents achieved humanlike intelligence for a week, allowing us to converse as equals with another species. Can you imagine the profound effect on humans to be able to "see" the world from another being's point of view? What would you do if God said, "I'll let you see the world through the eyes of any animal you choose." Which animal would you choose?

The Seeker Is a Finder

The emphasis of this book is on what aliens might look like and how their bodies might function. I also discuss the difficulties of recognizing an alien transmission from outer space—from creatures very different than ourselves. Chapter 7 gives several numerical and other test transmissions—just to show the difficulties we'd have determining the meaning behind alien transmissions. These transmissions allow you to fantasize you are sitting next to the famous Arecibo radio telescope and helping scientists decode enigmatic messages.[4] What problems would you encounter deciphering messages from beings on worlds shrouded eternally in clouds or from creatures that spend their time contemplating eternal philosophical truths and mathematical abstractions, with just a passing interest in talking with us? Would it be good for humanity to receive a transmission from an advanced extraterrestrial civilization? If we receive a transmission, should we reply?

In closing, let me remind you that humans are a moment in astronomic time, transient guests of Earth. Our minds have not sufficiently evolved to comprehend all the mysteries of outer space, of alien minds, and of alien races. Our brains, which evolved to enable us to run from cheetahs on the African grasslands, may not permit us to communicate with many forms of alien life or to understand their thought processes. Given this potential limitation, we hope and search for knowledge and understanding, always keeping in mind the aforementioned ancient Persian proverb:

"The seeker is a finder."

WHAT ALiENS LOOK LIKE

Aliens will not resemble anything we've seen. Considering that octopi, sea cucumbers, and oak trees are all very closely related to us, an alien visitor would look less like us than does a squid. Some fossils in the ancient Burgess shale are so alien we can't determine which end of the creatures is up, and yet these monsters evolved right here on Earth from the same origins as we did.

—Johan Forsberg

Other intelligent life-forms will differ greatly in appearance—they may resemble the creature in *E.T.* or startle us with their beauty—but life itself is common, I'm certain.

—Frank Drake

To consider the Earth as the only populated world in infinite space is as absurd as to assert that in an entire field of millet, only one grain will grow.

—Metrodorus, Greek philosopher of the fourth century B.C.

If the aliens look like Naomi Campbell, I'll welcome them with open arms.

—Marc Suzerain

Evolution

You are walking through the Nevada desert with Captain Steven Hiller, the hero of the science fiction movie *Independence Day*. Suddenly you hear a sound in the sky as a huge alien ship veiled in fiery clouds appears above you.

All over Earth alien craft launch an incredible attack. The alien destroyers are 15 miles (24 km) long, and the mother ship is 200 miles (322 km) in length, both impossible for any Earthly weaponry to destroy.

An alien appears before you (Figure 1.1). It must be a hoax. If the creature evolved on an alien world, why should it look so humanoid? The alien stands upright and is bilaterally symmetric; that is, its left and right sides look the same. It has fingers, two jointed legs and arms, a head with two eyes, and a large cranium.

Despite the alien's strange appearance, from a distance of 50 yards at dusk you might mistake it for human. But this seems somewhat unrealistic. In fact, stripped of its biomechanical armor, the alien looks more human than does an Earthly lemur, with whom we share more than 95 percent of our genetic material.

Science fiction writers have explored a far greater diversity of alien life-forms in books than Hollywood can ever explore in movies, because the Hollywood alien must trigger instantaneous emotional impact; this requires a design based on recognizable human facial expressions of threat and menace. In fact, most of the "evil" Hollywood aliens since 1953's *The War of the Worlds* have had a tendency to look mean and cranky, or like skull-faced sex fiends. In reality, if we ever meet real aliens we will have a hard time understanding their moods by looking at them.

The best way we can guess at how alien life might appear is to consider the evolution of animal shapes on Earth. The idea that alien evolution will lead to creatures that look like us is far-fetched—despite the fact that on *Star Trek*, Mr. Spock looks almost exactly like us even though he was born on planet Vulcan to a Vulcan father. Mr. Spock's mother was human, yet somehow his father, who hailed from an entirely different planet, was able to fertilize her—something less likely than your or my being able to mate with our close evolutionary cousins such as the octopuses and squids. Similarly, the aliens of Whitley Strieber's

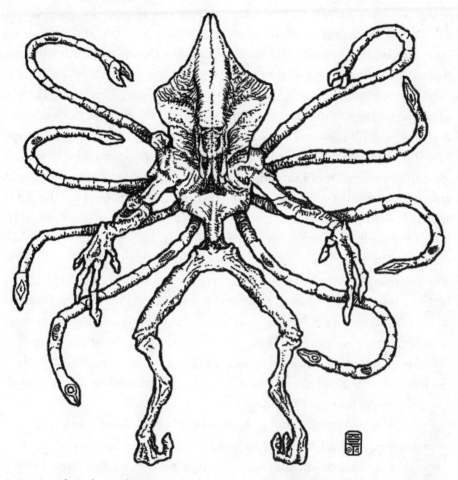

1.1 *An alien from the movie* Independence Day. *The alien stands upright and is bilaterally symmetric. It has fingers, two jointed legs and arms, and a head with two eyes. The extra snakelike appendages are part of its biomechanical armor. (Drawing by Ken DeVries.)*

Communion, and also those drawn by people claiming to have been abducted by aliens, have faces vaguely resembling our own. These creatures, like the aliens in Steven Spielberg's *Close Encounters of the Third Kind,* have large, smooth heads and huge black eyes. Again, they are also a little too human-looking, considering the quite different evolutionary pathways we'd expect on different worlds. Obviously, Hollywood production costs can be kept down if aliens are simply humans wearing sophisticated masks and makeup with dripping goo. Why is it that so many of the recent Hollywood extraterrestrials tend to be wet—could the slimy goo suggest amniotic fluid, mucus, evisceration, and dangerous body fluids? Perhaps all the alien drool reminds us of rabid animals and therefore is something to be feared.

In our real universe, there are many reasons why it is unlikely that aliens would take human form. For one thing, the diverse rates and directions of evolution on Earth and the fact that many types of creatures have become extinct show that there is no goal-directed route from single-cell organisms to an intelligent human. Given only slightly different starting conditions on Earth, humans would not have evolved. In other words, evolution is so sensitive to small changes that if we were to rewind and play back the "tape" of evolution, and raise the Earth's initial overall temperature by just a degree, humankind would not exist. The enormous diversity of life today represents only a small fraction of what is possible. Moreover, if humans were wiped out today, humans would not arise again. This means that on another world, the same genetic systems and genes will not arise. This also suggests that finding another planet with humans, dinosaurs, or apes is more unlikely than finding an island off the coast of New Jersey where the natives speak English through snouts like dogs'.

Evolution on Earth tells us a lot about possible alien shapes. Although every detail must be different, there are patterns of general problems, and common solutions to those problems, that would apply to life on alien worlds. In the course of Earth's history, whenever life-forms have had a problem to solve, they have solved it in remarkably similar ways. For example, three very unrelated animals—a dolphin (a mammal), a salmon (a fish), and an ichthyosaur[1] (an extinct reptile)—all have swum in coastal waters darting about in search of small fish to eat. These three creatures have very little to do with one another biochemically, genetically, or evolutionarily, yet they all have a similar look. On first glance, they look like nothing more than living, breathing, torpedoes. Although they are biologically quite different, they have all evolved streamlined bodies to help them travel quickly through the water. This is an example of *convergent evolution,* and we might expect aquatic aliens that feed on smaller, quick-moving aliens also to have similarly streamlined bodies.

With convergent evolution, successful solutions arise independently in different animal lines separated in time and place. The reason for the similarity of solutions is clear: Animals encounter similar environmental problems and cope with them in a similar way because that solution is an efficient one. My favorite examples are prehensile (grasping) tails that have developed in all of the following: opposums, chameleons, sea horses, binturongs (catlike carnivores found in the dense forests of southern Asia), kinkajous (Central American raccoons), pangolins and tamanduas (anteaters), South American tree porcupines, *Aneides* (a genus of tree-dwelling amphibians), phalangers (Australasian marsupials), and some monkeys. Similar needs in these very different animals have evoked a similar response in the structure and function of the tail.[2] These universal solutions will be found on other planets with life.

The only chance of finding aliens that look exactly like us is in parallel universes—worlds that resemble our own and perhaps even occupy the same space as our own in some ghostly manner. Although the multiple-universe concept may seem far-fetched, serious physicists have considered such a possibility. For example, the physicist John Wheeler of Princeton University proposed that multiple universes were spawned during the Big Bang that created our universe. Most of these universes have laws of physics different from ours and would probably be impossible to observe. Also, Hugh Everett III's doctoral thesis "Relative State Formulation of Quantum Mechanics" (reprinted in the July 1957 issue of *Reviews of Modern Physics*) outlines a controversial theory according to which the universe at every instant branches into countless parallel worlds; this is called the "many-worlds" interpretation of quantum mechanics. However, human consciousness works in such a way that it is only aware of one universe at a time. The "many-worlds" theory holds that whenever the universe is confronted by a choice of paths at the quantum level, it actually follows both possibilities, splitting into two universes. These universes are often described as "parallel worlds," although, mathematically speaking, they are orthogonal or at right angles to each other. In the many-worlds theory, there are an infinite number of universes; if this is true, then all kinds of eerie worlds exits where virtually everything must be true. There is a universe where fairy tales are true: A real Dorothy lives in Kansas dreaming about the Wizard of Oz; a real Adam and Eve live in a Garden of Eden; and alien abductions really do occur all the time. The theory also implies the existence of infinite universes so strange we could not describe them. My favorite tales of parallel worlds are those of Robert Heinlein. In his science fiction novel *The Number of the Beast* there is a parallel world similar to ours except that the letter *J* does not appear in the English language. Luckily, the protagonists in the book have a device that lets them explore parallel worlds from the safety of their high-tech vehicle. In contrast, the protagonist in Heinlein's novel *Job* shifts through parallel worlds without control. Just as he makes some money in one America, he shifts to a slightly different America where his money is no longer valid currency, which tends to make his life miserable.

The many-worlds theory suggests that a being existing outside of spacetime would see all conceivable forks, all possible four-dimensional space times, as always having existed.[3] How could a being deal with such knowledge and not become insane? A god would see all Earths: those where no inhabitants believe in God, those where all inhabitants believe in God, and everything in between. According to the many-worlds theory, there would be universes where Jesus was the son of God, universes where Jesus was the son of the devil, and universes where Jesus did not exist.

A large portion of Everett's many-worlds interpretation is concerned with events on the submicroscopic level. For example, the theory predicts that every time an electron either moves or fails to move to a new energy level, a new universe is created. Currently the degree to which quantum (submicroscopic) theories impact reality at the macroscopic, human level is not clear.

Alien Symmetries

It is certainly possible for an alien body to possess internal organs similar to those of its Earthly counterparts because alien bodies will have to perform functions that are carried out most efficiently by specialized tissues. For example, aliens may have digestive and excretory systems, a transport system to distribute nutrients through the body, and specialized organs to facilitate movement. Evolutionary pressure would probably lead to such familiar ecological classes and phenomena as carnivores, herbivores, parasites, and beneficial symbiotic relationships. Aliens with any technological capacity will have appendages comparable to hands and feet for manipulating objects. Technological aliens must also have senses, such as sight, touch, or hearing, although the precise nature of the senses that evolve on another world would depend on the environment. For example, some aliens may have eyes sensitive in the infrared or ultraviolet regions of the spectrum because this sensitivity has survival value on a particular world. Creatures manifesting some of these basic development trends would be quite different from us, with various possible symmetries, and they could be as big as *Tyrannosaurus* or as small as a mouse, depending on gravity and other factors.

To a biologist, *symmetry* refers to the orderly repetition of the parts in an animal or plant. Often symmetry refers to the position of body parts on opposite sides of a dividing line or distributed around a central point or axis. Some of the most successful life-forms on Earth have *bilateral symmetry,* meaning that only one plane of symmetry divides an animal into symmetrical halves. (For example, it's possible to slice a human with a vertical cut and get two similar pieces.) Bilateral symmetry is characteristic of the vast majority of animals, including insects, fishes, amphibians, reptiles, birds, mammals, and most crustaceans (Figure 1.2).

It appears that our ancestors were aquatic animals. If aliens also evolved from aquatic species, they too may have bilateral symmetry because this is an efficient way to produce a streamlined, muscular body for catching food and fleeing from predators in the water—especially when compared to slower life-forms with *radial symmetry,* like the more sedentary starfish, urchins, and jellyfish (see Figure 1.3). In radial symmetry, the body has the general form of a cylinder or bowl with a central axis from which the body parts radiate, or along which they are arranged in regular fashion.

From *Star Trek, The X-Files, Independence Day,* and *Mars Attacks* to *Men in Black*, most TV and Hollywood science fiction authors have created bilaterally symmetric creatures for their stories. However, occasionally science fiction authors have suggested radial forms that are quite interesting to contemplate. For example, Naomi Mitchison's

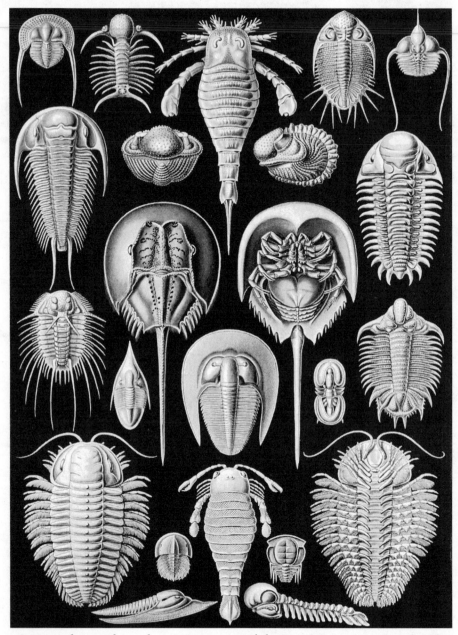

1.2 Horseshoe crabs and various species of their extinct ancestors (trilobites), all of which exhibit bilateral symmetry.

1.3 Various species of Semaeostomeae (an order of jellyfishes) that exhibit radial symmetry.

Memoirs of a Spacewoman describes "Radiates," intelligent five-armed creatures resembling starfish (Figure 1.4) that live in villages composed of long, low buildings with ceilings decorated with fungi that grow in spiral patterns. Radiates don't think in terms of dualities, having instead a five-valued system of logic.

Another fascinating radially symmetric creature from science fiction is the Abyormenite from Hal Clement's *Cycle of Fire* (Figure 1.5). These six-tentacled creatures live on a planet orbiting a complex two-sun system that creates oscillating cold and dark weather cycles. Abyormenites live during the 65-year period when their planet has a warm temperature. When their planet enters its 65-year cold period, all Abyormenites die, leaving spores in the bodies of intelligent beings who dominate the planet in the cold years. When the hot years return, the cold life in turn dies, depositing its reproductive spores in the bodies of the next generation of Abyormenites. Since the races require one another for their existence, they have agreed voluntarily to the oscillating

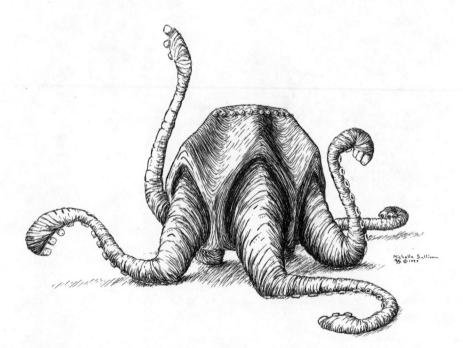

1.4 Radiates from Naomi Mitchison's novel Memoirs of a Spacewoman. *(Drawing by Michelle Sullivan.)*

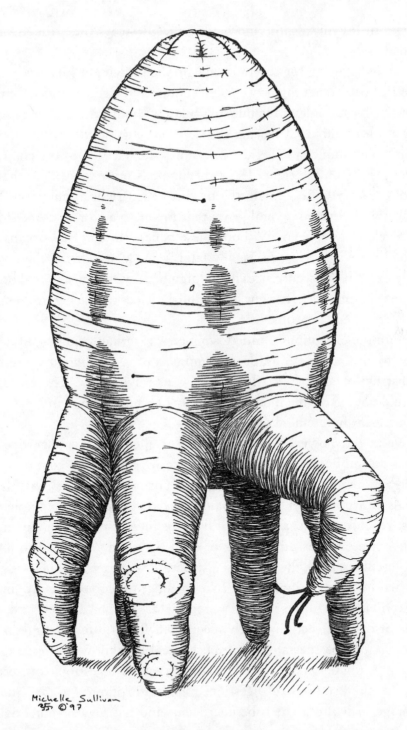

1.5 *Abyormenite from Hal Clement's* Cycle of Fire. *(Drawing by Michelle Sullivan.)*

cycles of life and death, and neither race destroys the artifacts of the other.

Let us propose for the moment an alien having a bilaterally symmetric body. What might we guess about the structure of its organ systems? On each side of the human's bilateral body is an eye, ear, nostril, nipple, leg, and arm. Beneath the skin, our guts do not exhibit this remarkable symmetry. The heart occupies the left side of the chest; the liver resides on the right. The right lung has more lobes than the left. Biologists trying to explain the origins of left-right asymmetries have recently discovered several genes that prefer to act on just one side of a developing embryo. Without these genes, the internal organs and blood vessels go awry in usually fatal ways. Mutations in these genes help explain the occurrences of children born with their internal organs inverted along the left-right axis, a birth defect that generates remarkably few medical problems.[4]

With the exception of radial symmetry, external form has little relation to internal anatomy, because animals of very different anatomical construction may have the same type of symmetry. Therefore, we would not expect the internal anatomy of a bilaterally symmetric alien to necessarily resemble our own.

We might, however, expect intelligent aliens to have digestive systems resembling a tube structure, since we see this so commonly on Earth in many different environments. For example, most Earthly animals above the level of cnidarians and flatworms have a complete digestive tract, that is, a tube with two openings, a mouth and an anus.[5] There are obvious advantages of such a system compared to a gastrovascular cavity, the pouchlike structure with one opening found in flatworms. For example, with two openings the food can move in one direction through the tubular system, which can be divided into a series of distinct sections, each specialized for a different function. A section may be specialized for mechanical breakdown of large pieces of food, for temporary storage, for enzymatic digestion, for absorption of the products of digestion, for reabsorption of water, and for storage of wastes. The tube is efficient and has greater potential for special evolutionary modifications useful for different environments and foods.

Alien Brains

Stop for a moment and imagine yourself in an intergalactic "zoo" of living brains in bottles. You walk down fluorescent-lit corridors filled with gray, wrinkled things stored in fluid-filled jars.

You give a little tap on the jar marked "Alpha Centauri," and a creature's cerebrum jiggles like a nervous mango. You put "Alpha Centauri" back in its place. On your right are a few clear jars. You reach for the one marked "Tau Ceti," open it, and drag your fingers lingeringly over the gray-white frontal lobes.

What might an alien brain really look like? Would aliens even have brains and nervous systems? On Earth, a creature's nervous system is made up of an organized group of cells specialized for sending signals from sensory receptors through a nerve cell network. The nervous system enables an animal to respond to changes in its external and internal environment.

Nervous systems are of two general types: diffuse and centralized. A noncentralized nervous system, known as a *diffuse nervous system,* is found only in lower invertebrates (animals without backbones), particularly among coelenterates like jellyfishes and hydras, which are radially symmetrical. In a diffuse-type system, there is no brain, and the nerve cells are distributed throughout the organism in a netlike pattern. However, most other animals that have a nervous system show some degree of centralization, meaning that there is some part that coordinates information and directs responses. Even the radially symmetrical echinoderms (starfishes) have a dominant central nerve ring from which extend radial nerves. Echinoderms, however, have no brain.

Judging from Earthly creatures, intelligent alien life-forms would have some kind of brain or central region for interpreting signals coming from the environment. We've already talked about bilateral symmetry, the body plan that characterizes most higher invertebrates and all vertebrates. A central nervous system with a brain seems to have been an evolutionary outgrowth of bilateral symmetry. From an evolutionary standpoint, certain flatworms are the most primitive animals with a central nervous system, though their brain is little more than a small bulging of the nerve cords in the head region of the animal. A more complex central nervous system is found among the annelids (earthworms and leeches) and the arthropods (insects and crustaceans), which have an unmistakable brain and ventral nerve cords. ("Ventral" means located near the lower belly surface of an animal, opposite the back.) Centralization reigns supreme in the vertebrates, which have a well-developed brain and a dorsal nerve cord (the spinal cord).

If you ever stumble across a dead alien and find it has a nervous system, you can assume that the alien moved around and interacted strongly with its environment. It seems that a precondition for the development of a complex nervous system is an active, mobile, or predatory lifestyle with appendages for manipulating the environment. Intelligent aliens may evolve from carnivores because intelligence has survival value in cooperative hunting. In such aliens, the central controlling brain would be close to the primary sense organs, so that the nerves connecting the sense organs are short and correspondingly fast. Such an alien must also have sensing organs at the front of the body near the mouth. If the alien must chemically sense the food before eating it, a noselike organ for this sense must be located near the mouth. Paired sense organs may be evolutionarily valuable for producing such characteristics as binocular vision or sound localization. Similarly, it would not surprise me if aliens are bilaterally symmetric and have large ganglia of nerves near the front of the head and near the primary sense organs, such as those exhibited by the alien in the movie *Independence Day*.

Aliens will probably have heads to house a brainlike organ near the primary sense organs. On Earth we see the evolutionary beginning of this cephalization (concentration of nervous functions in the head region of an animal) in invertebrates. The highest degree of development of the invertebrate nervous system is attained by the cephalopods (squids, cuttlefishes, and octopuses) among the mollusks and by the insects and spiders among the arthropods. Control and coordination of specific functions, such as locomotion and feeding, are compartmentalized in particular parts of the nervous system. The complex nervous system of the cephalopods is correlated with the active movement and predatory habits of these animals.

The need for a well-protected area to house the center of the nervous system does not *have* to lead to a brain in a head of an alien, as is the case of Earthly organisms. For example, I can imagine that an alien brain might be housed within the protective surroundings of the alien's trunk; after all, it's good enough protection for the digestive and circulatory system. Sensory organs should be nearby.

Many aspects of the brain's structure are related to overall alien body structure—a topic of significant interest to science fiction writers. One large determinant of alien body and brain structure is gravity. If the force of gravity were less, organisms might grow larger or be more delicately constructed. On high-gravity worlds, bones and muscles would have to be much stronger. Presumably, alien brains on these worlds would be quite compact. On Earth, our main contenders for intelligence include the mammalian order Cetacea—whales and dolphins. Because these animals live in the sea, which supports their weight, they can develop large brains and heavy heads—something quite difficult on land unless an animal stands upright.

Would brains capable of high intelligence evolve on other worlds? Almost all science fiction writers overlook the fact that intelligence is not evolutionarily selected for on Earth. For example, the vast majority of Earthly organisms that are highly successful from an evolutionary standpoint do not have high intelligence, and these animals would not have much additional success if they became more intelligent. Our own intelligence will not guarantee us perpetual dominion over the worms, beetles, and flies. From an evolutionary standpoint, a cosmic gambler would not bet on humans' having a longer evolutionary lifetime than the ants.

How would an alien brain affect alien behavior? Intelligent alien beings would be highly communicative and have complex social relationships—if they follow the trends exhibited by life on Earth. For example, birds are more intelligent than reptiles and are also more communicative. Social relations in birds are more complex than those among reptiles, particularly when it comes to rearing young. The *Time* magazine writer Nancy Gibbs notes that on Earth the emotion of love is made possible through the neocortex, an evolutionarily advanced part of the brain enabling humans to plan, learn, and remember. Reptiles have no neocortex and cannot truly experience maternal love, and this is why many reptile babies, such as baby snakes, hide to avoid being eaten by their parents. The more connections there are between the limbic system (a primitive brain part that is the site of delight, disgust, fear, and anger) and the neocortex, the more emotional responses are possible.

On Arms, Legs, and Octopuses

For every sense that is vital to an animal, there is an accompanying, noticeable accoutrement on the body surface. For example, if hearing is vital for an alien, I'd expect some form of ear that may be able to swivel in the direction of sound. If the sense of smell is important, we would expect a nose, long proboscis, or forward-projecting snout. On the other hand, if these senses are diminished or unimportant, and an alien relies on a sense like the electrical sense of fish, we would expect the external accoutrements—eyes, ears, and noses—to be smaller.

Life-forms respond to the available input signals around them. This means that many aliens would have organs for sight and sound, because the universe is often bathed in light and sound. Aliens would also have some sense of touch so that they could respond to the physical world by moving around objects, avoiding dangerously sharp shapes, and so forth. The tactile sense is probably the most basic of all senses, and there are very few life-forms on Earth, however simple, that do not respond in some way to being touched. *Intelligent* aliens would have additional senses; it seems unlikely that a complex and intelligent creature could evolve with one sense alone.[6] Additional senses are needed to confirm and refine an intelligent creature's perception of the immediate environment. This means that other senses are likely to accompany the sense of touch.

Would an alien have arms and legs? The alien in *Independence Day* has two jointed legs, which is quite reasonable. Again, on Earth, many very different organisms have evolved jointed legs for efficient locomotion over different kinds of terrain. A much larger number of legs might make for difficulties in coordination and slowness in movement, while an odd number of legs could conceivably create an awkward imbalance. Therefore, the fastest runners would probably have only a small number of legs in pairs. Although humans have two legs and two arms, it may be more efficient to have more limbs for running and manipulating the environment. For example, four-legged animals are faster than two-legged ones.

1.6   The highly intelligent octopus, a perfect model for an alien lifeform. The octopus has senses we can hardly fathom and a weird brain that wraps around its esophagus.

Another model for an alien body is the octopus, which has sensitive tentacles at one end of the body, and a mouth at the base of the tentacles (Figure 1.6). Octopuses' eyes and electric organs are on the body surface. (Electrical organs are discussed in detail in Chapter 2.) The octopus is a very intelligent creature—some say as smart as a dog—with a multitude of complex behaviors such as the ability to learn from watching others of their kind.

Octopuses are truly alien; they can see polarized light, which we cannot.[7] They also have highly developed senses of touch, taste, and smell, and an organ for a sense that can best be described as one for hearing: fine hairs along the head and arms that can detect disturbances up to 98 feet (30 m) away.

The brains of octopuses are weird! The highly textured brain is a tightly packed mass of lobes that lies between the eyes and that encircles the esophagus. This type of brain has its drawbacks—researchers have discovered spines lodged in octopuses' brains, the result of a meal going down the wrong way. Today scientists wonder why such a creature needs such a large brain. We do know that the octopus, the most intelligent of all invertebrates, has highly developed pigment-bearing cells and can rapidly change its skin color. To avoid being eaten, the octopuses and their kin can blend in with their surroundings by positioning their tentacles to mimic floating sea grass, or by flexing skin muscles that can change the texture of their skins. They can adopt patterns that range from full-body speckles to dramatic black and white tiger stripes. The more distinctive patterns are used as signals during courtship, hunting, and male-to-male aggressive encounters, and in response to a threat.

I wonder why octopuses are so intelligent? After all, they are short-lived, usually solitary creatures that sometimes meet other octopuses only once, to copulate. Their brains evolved entirely separately from the brains of vertebrates, and they have an entirely different design. Could they also house an "alien" form of intelligence? Because octopuses' behavior and brains are so unusual, we might be overlooking their greatest cognitive feats.

As might be expected of a creature with such a complex nervous system, the octopus is very emotional. It can't hide its emotions from us, since the emotions are expressed by color changes. From its normal light-brownish color, an octopus may blanch to white or turn through stages of pink to red as it expresses fear, anger, irritation, excitement, or other feelings. If aliens exhibited similar vivid changes to their coloration in response to emotion, would they be hesitant to interact with us because they would reveal their inner feelings in any negotiations? How would our own society change if our emotions were more obvious to others?

The fact that the octopus has no skeleton makes it quite a versatile creature. Octopuses can stretch themselves thin, like rubber, and turn their eyes obliquely, so that even a huge octopus can slide through barely visible cracks and openings. They can lengthen their arms by stretching them. Even the head can become quite thin, allowing them to escape from standard aquaria when in captivity.

Octopuslike aliens are certainly possible. However, if there are intelligent, *technologically advanced* life-forms on other worlds, I think it is more likely that they dwell on the land than in the water. Water dwellers would have great difficulty using tools in a viscous, turbulent medium. For example, try to assemble a Swiss watch with tiny gears or other small parts underwater.

It's also doubtful that technological aliens would fly in the air like birds. Birdlike aliens are not likely to develop high intelligence because they must be light. They can't afford the weight of a large brain nor of a large heartlike organ needed to supply the brain with sustenance. Of course, it might be possible to have intelligent flying aliens on planets with lower gravities than that of Earth or with denser atmospheres.

Lessons from Science Fiction

Let's focus our attention on hypothetical aliens in the science fiction literature, because they raise interesting questions about language, culture, and even sexuality. I'm not too interested in aliens in the earliest science fiction literature because these were often the least realistic. For example, the early, evil aliens often had forms exactly like reptiles, insects, or spiders but were able to stand upright and were about human height. Many times the good aliens were mammalian, birdlike, or angelic. Even when aliens were quite strange in appearance, their motivations and mental worlds were all too understandable and humanlike.

In previous sections, we've mentioned aliens in modern movies like those in *Independence Day* and *Close Encounters of the Third Kind*, but there are much more imaginatively shaped creatures in science fiction novels. For example, the Cygnans, though still somewhat humanoid, are a race of intelligent creatures described in Donald Moffitt's novel *The Jupiter Theft*. These human-sized beings live on a giant gas planet orbiting a binary star system. When one of the stars collapses into a black hole, the Cygnans migrate from their world using 30-mile-long spaceships, the interiors of which contain huge, artificial forests where Cygnans live. Eventually, the Cygnans enter our Solar System and use pieces of Jupiter as a source of energy. Unfortunately, humans have great difficulty stopping the Cygnans. In fact, Cygnans have lost all interest in anything unrelated to survival, owing to their long isolation in their spaceships. They therefore have a total disregard for humans.

Figure 1.7 shows a typical Cygnan female with two legs and six "arms." In practice, the six arms can be used as either arms or legs. Sometimes Cygnans stand erect on their hind legs, tails hanging straight down. Sometimes they are on four legs, their torsos upright so that they are shaped like low-slung centaurs. A three-petaled tail folds to conceal sexual organs. The skeleton of the Cygnan is cartilaginous, like a shark's skeleton. The Cygnan's brain is located between the upper pair of limbs at the top of a spinal cord. The three eyes are placed on stalks in an equilateral triangle around a broad, flexible

1.7 *A Cygnan from* The Jupiter Theft. *(Drawing by Michelle Sullivan.)*

mouth. Inside the mouth is a harsh, rasping plate and a spiked, tubular tongue.

When a spacecraft from Earth travels to meet the Cygnan ship, the Cygnans capture Tod Jameson, and first contact begins:

> Jameson squinted at the nearest alien. It squinted back at him with its three stalked eyes. . . . There was something primitive

about the tapering, arrowhead-shaped skull. The jaws split it down the middle in a permanent reptilian smile.

The Cygnan's nervous system functions quite efficiently with synaptic reflexes much faster than those of a human being. If you were to hear a Cygnan talk, you would hear music—chords produced by several larynxes and the spiked, tubular tongue. Their speech depends mostly on pitch, and their language is very rich and varied, with more than a million phonemes—a phoneme being the smallest unit of sound in a language that can differentiate one word from another. (As a comparison, American English has 13 distinct vowel phonemes.) Jameson, the novel's protagonist, realizes that Cygnan language is musical when he first hears chords coming from their mouths:

> There was a sound like a maniac trying to play Bartók on the harmonica, and Jameson realized it had been made by one of the Cygnans. The other Cygnan answered with an incredibly rapid fragment of twelve-tone sounds. Jameson came to full attention. There had been chords in all that quick passage-work, transitory but unmistakable, as if the Cygnan possessed multiple larynxes.

Given time, Jameson gradually comes to understand the Cygnan language:

> The large Cygnan turned to him again and made a sharp attention-getting sound. Then it touched itself on the mouth and the tip of its petalled tail and sounded the tetrachord again. It waited. Jameson hesitated. The tetrachord had been easy. It was a handy, one-phoneme identification. Like, Jameson thought, a human saying "I." But this was more complicated. The second Cygnan repeated it for him until he got it straight.

If you look closely at Figure 1.7, you'll see a small oval shape at the bottom front. This is the Cygnan male, a parasitic organism that remains with the female for life. At the head of the male is a feeding tube, which can extend into the female. If this sexual arrangement seems unrealistic, consider my favorite Earthly example of parasitism that shows the effect

1.8 Spiders often exhibit vast size differences between males and females of the same species. Shown here is a silk spider from Malaysia of the genus Nephila *(male above, female below).*

1.9 *The anglerfish.*

of environment on sex determinism: the sea worm *Bonnelia*. If the free-swimming *Bonnelia* larvae settle on the sea bottom, they develop into females, each with a long proboscis (flexible tube). On the other hand, larvae that land on the female proboscis develop into tiny males that lack digestive organs and exist in parasitic fashion in the genital ducts of the female. (When I once lectured about this creature, one male chauvinist in the audience remarked that this was the ultimate example of women's liberation in the animal kingdom.)

Also consider size differences in black widow and other spiders (Figure 1.8), where the female is much larger than her mate and the courtship may prove fatal to the male. In fact, the male is seldom seen because it is one fourth the size of the female and because it is often eaten by the female after mating. There are other numerous instances of large size differences in the animal kingdom. For example, the male deep-sea anglerfish is much smaller than the female and lives parasitically attached to her (Figure 1.9). Like some nightmarish "fatal attraction," his mouth fuses with her skin, and the bloodstreams of the two fish become connected, the male thereafter remaining totally dependent on the female for nourishment. What would your reaction be to alien life-forms exhibiting such interdependence of the sexes? Depending on your emotional perspective (and perhaps *their* emotional perspective), you might consider either the alien couple as the ultimate form of spiritual bonding, or the ultimate form of enslavement. I'll talk further about hypothetical alien sexuality in Chapter 6.

Do you think intelligent, technological aliens would be even less humanoid than a Cygnan? My favorite nonhumanoid alien in science fiction is the Cryer from Joseph Green's *Conscience Interplanetary*. The Cryer lives on planet Crystal, which has an oxygen atmosphere. The Cryer resembles a human-sized bush with a trunk made of crystal and metal (Figure 1.10), and the creature's leaves are made of sharp glass. A Cryer is actually a single unit of a planetwide silicon-based plant intelligence. An Earthly analogy is the ant, an independent entity that forms a colony with characteristics of a group intelligence. Ants are based on carbon, but on planet Crystal all life is based on silicon and metallic elements.

The trunk of a Cryer contains silicon memory units powered by a low-voltage solar storage battery and connected by fine silver wires.

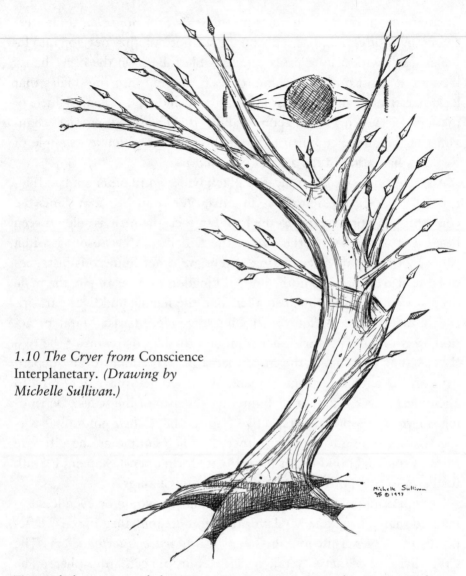

1.10 The Cryer from Conscience Interplanetary. *(Drawing by Michelle Sullivan.)*

Toward the center of the Cryer is an organic air-vibration membrane created for it by the planetwide intelligence so that the Cryer can speak with human beings. The dish-shaped membrane is held in place by wires, like a spider web, so that the dish can act like a vibrating speaker. A magnetic field generated by silver wire coils hanging on both sides of the speaker cause the speaker to vibrate and produce sound.

The most fascinating aspect of Cryers is how they communicate with one another. Beneath the surface of planet Crystal is a nervous system of fine silver wire that interconnects thousands of smaller Cryerlike units forming the planetwide intelligence called Unity. Each unit has a specialized function—for example, some store electricity generated by sunlight, some extract silver for constructing the nervous system, some provide memory storage, and some act as sensor units. The overall intelligence is able to perceive motion, temperature, position, vibrations, and electrical potential through its member units. Unity and its subunits sleep while absorbing sunlight.

If you think the idea of a hive mind like Unity is implausible, consider the behavior of ants and bees, which is very complex by insect standards, and their sophisticated processes of communication, including chemical messages and dances signaling the location of food sources. These insects form a group intelligence that is smarter than any one of the components.

Another example of an Earthly hive mind consists of termites that build huge hills ventilated by a complex system of passages. These passages cleanse the mound of excessive carbon dioxide and take in fresh oxygen. Worker termites supervise and correct the ventilation system, alternately narrowing the vents or widening them to adjust the temperature and oxygen levels. The amazing cooperation between workers highlights the group intelligence: Although the individuals are cold-blooded, they can work together to regulate temperature in the hive just as a warm-blooded creature does. The colony behaves like the body of an organism. The group brain produces intelligent solutions, although the individuals are capable only of carrying out stereotyped actions.

What would it be like to interact with an alien group brain that possessed an aggregate intelligence smarter than humans'? Even more mind-boggling is the possibility of a "superbrain," where each individual is highly intelligent and the aggregate is superintelligent. (A crude example of this is a team of smart engineers building a car: Each engineer knows only about a specific part of the car, but the engineers come together to form a group intelligence capable of creating an object that no individual can create.) In *Other Senses, Other Worlds,* Doris and David Jonas ask,

> What is the "other way of knowing," by which the termites "know" what they have to do and when they have to do it? In-

structions cannot be brought to them quickly enough by messengers, since the distances within the hill are far too great. There is no perceptible means of communication that we can discover. Not rarely in nature a group brain functions as an instrument for decision-making in a way startlingly like an intelligent individual brain.

Indeed, the termite "civilization" is among the most impressive in the animal kingdom. The termites' monolithic cities soar 18 feet (5.5 m) high, representing the greatest modification of the natural landscape caused by multicellular animals (aside from man). If a termite were the size of a man, a termite hill would be 4,000 feet (1,223 m) high, three times the height of the Empire State Building—and termites don't use special tools to help them create.

The group mind of termites is like a machine for forming statistical assessments after sensing the surrounding environment. Perhaps termites find answers to their lives' questions in the same way a computer, given a variety of input signals, determines statistical outcomes, or the way that our brains' neurons collectively come to decisions as a result of chemically weighting a variety of input signals. In the same way that a mountain "knows" when to produce an avalanche, given certain stimuli, the group termite mind knows when to build new hills or tunnels.

The termite's decision is not an attribute of anything that exists physically in the termite group; rather, it is an *emergent quality* of the complexity of interactions between the multitude of individuals, in many ways like a thought—a product of complex interactions of millions of nerve cells. The termites remind me of "self-organizing systems," in which large-scale patterns arise from simple rules operating on tiny components of a system. Examples of such behaviors arise in traffic jams, the aggregations of slime molds or bacteria, and the flocking of birds.

Aliens, like termites, might also possess a group mind that emerges as a function of the complexity of their group life. Depending on the mechanism of communication between members, the individuals of such a group mind could be spread out over miles of territory. On Earth, animal colonies vary tremendously in geographical size. For example, the largest animal colony ever discovered was a colony of black-tailed prairie dogs, rodents found in the western United States. They typically build huge colonies, and a single colony discovered in

1901 contained about 400 million individuals and was estimated to cover 24,000 square miles (62,000 square km).

Of course, alien colonies could be totally unlike anything on Earth, with completely different individual and group senses. If we were to meet an alien colony, communication with individuals might be impossible. Each individual might not even be cognizant of the overall group mind and pattern. Moreover, there does not even have to be a leader overseeing the group mind. The mind is simply the aggregate behavior and response of the ensemble of individuals.

A complex group mind arising from simpler individuals reminds me of the Borg from the TV and movie series *Star Trek*. Individual Borg units (Figure 1.11) are both organic and mechanical, and equipped with various hardware for specified tasks. Borg units form a collective intelligence or consciousness, each unit communicating through a network of "subspace" links giving them the ability to learn and adapt instantly. The Borg's "prime objective" is to assimilate any creatures or raw materials with which it comes in contact.

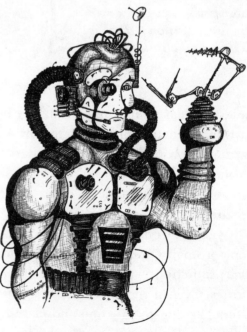

I wonder how the Borg may have evolved. *Star Trek* writers tell us that the Borg are a race of cybernetically enhanced humanoids. The Borg drones grow as normal humanoids in the first period of their life and then are implanted with cybernetic parts and finally placed into the collective. Each Borg has a different device on its body enabling it to perform a specific function in the Borg Collective, such as defense, communication, and navigation. As a result of efficient communication, Borg units have lost their individuality and the whole collective acts

1.11 A Borg from Star Trek *forms part of a collective consciousness. (Drawing by Brian Mansfield.)*

1.12 *The highly organized structures of beehives and hornet nests have been used by science fiction writers as a model for alien societies.*

and thinks as a single organism. The Borg thoughts are quick, their reaction speed alarming. Despite how alien this may seem, we see rapid group-mind responses here on Earth when ants are suddenly confronted with drops of rain or with an invader, and the ant collective responds with incredible swiftness. The same kind of group response occurs in bee hives and hornet nests (Figure 1.12).

Cygnans, Cryers, Borgs, and various kinds of group minds represent only a small sampling of possible alien biologies. One reason I enjoy science fiction is that it provides so many possibilities for alien life-forms. Even if the aliens are not entirely plausible, at the least they stimulate our imagination and permit us to discuss the pros and cons of certain hypotheses on the physical forms of extraterrestrials. Going beyond alien *form,* it is more difficult to hypothesize about how an alien might *behave.* However, to a large extent, our own behaviors are biologically determined, and this would be so for any alien life-forms.

I would expect most intelligent (non-group-mind) aliens to have young requiring a long learning period, promoted by slow growth and prolonged development. This fostered attachment behavior between caring adults and their offspring might even lead to morals similar as our own.

For those of you interested in the incredible range of alien physiologies in science fiction, take a look at James White's *Sector General* series, which describes an interspecies hospital on the rim of the Galaxy. His fascinating cast of characters, and his belief in the basic decency of all intelligent life-forms, pervade his stories. In *Major Operation* White describes various categories of aliens:

> We have evolved a four-letter classification system for our incoming extraterrestrial patients. The first letter denotes the level of physical evolution. . . . The second indicates the type and distribution of limbs and sense organs, and the other two the combination of metabolism and gravity-pressure requirements, which in turn gives an indication of the physical mass and form of tegument possessed by a being. Usually we have to remind some of our E-T students at this point that the initial letter of their classification should not be allowed to give them feelings of inferiority, and that the level of physical evolution has no relation to the level of intelligence.

In White's books, species designated by the prefix A, B, or C are water breathers. According to White's theories, on most worlds life began in the seas, and these water breathers evolved high intelligence without having to leave the water. D through F are warm-blooded oxygen breathers, "into which group fell most of the intelligent races of the galaxy." The G to K types are also oxygen breathing but insectile. The L's and M's are light-gravity, winged beings.

Continuing with White's imaginative classification scheme, chlorine-breathing life-forms are classified as O and P groups, and subsequent letters specify more bizarre organisms: frigid-blooded or crystalline beings, radiation eaters, aliens that modify their physical structure at will, and creatures that are very highly evolved. The prefix V denotes those creatures possessing extrasensory powers developed to such an extent as to make walking or manipulatory limbs unnecessary. White discusses his classification system in detail:

> "There are anomalies in the system . . . but those can be blamed on a lack of imagination by its originator. The AACP life-form, for instance, has a vegetable metabolism. Normally the prefix A denotes a water breather, there being nothing lower in the system than the piscatorial life forms, but the AACPs are intelligent vegetables, and plants came before fish.

Jack Chalker is my favorite science fiction writer for his sheer diversity of aliens. In his *Well of Souls* series, Chalker describes a world consisting of a hexagonal grid with 1,560 hexagonal areas, each with different creatures. In Chalker's books, the protagonist wonders whether the various intelligent Well-World races can mingle and socialize with one another. He is told that the life-forms in each hex are so very different, that socializing becomes difficult: "Could you really be good buddies with a nine-foot-tall hairy spider that ate live flesh, even if it also played chess and loved orchestral music?"

The Well-World's Northern Hemisphere contains 780 non-carbon-based life-forms with various ambassadors. The North Zone is an alien nightmare beyond imagination. Some of the life-forms are so wildly alien that they cannot find common ground with one another. Their utterances often don't make sense; their frame of reference and concepts are too alien to permit much communication. Races of the Northern Hemisphere include:

1. Astilgol—Symbiotic creatures resembling a set of hanging crystal chimes on which is set an invisible bowl with little flashing lights. Silicon eaters.
2. Bozog—Sticky creatures that resemble two eggs sunny side up, full of gritty little balls and with cilia beneath. They can form their liquid-filled bodies into tentacles and can stick to walls.
3. Cuzicol—Nocturnal metallic yellow flowers with hundreds of sharp spikes. They stand on two spindly legs.
4. Masjenada—Blown-glass swans without heads or feet. They can combine and alter their body material. They have the annoying habit of flying through each other with no ill effect.

5. Uchjin—Nocturnal creatures resembling smears of paint dripping in midair. They range in color through the entire visible spectrum and live underground during daylight hours, pouring out from cracks in the ground at night. The atmosphere of their hex is mostly helium.

Despite these bizarre appearances, in principle these aliens can live in peace with one another. Yet, I wonder what would happen if humans encountered aliens that were so horrible-looking that we had difficulty controlling our revulsion to their appearance? Revulsion is less a matter of logic than a physiological reaction. Yes, we can learn to control our reactions and become sensitized—just as a medical doctor becomes accustomed to the normally horrifying sight of cadavers, slimy intestines, and gut-wrenching horror—but ugly aliens would place a strain on alien-human interactions. It would take time to overcome; prejudices are slow to break down. By the same token, if we are horrified by a particular alien appearance, it is possible that the alien will feel horrified about our own looks. Our relations with aliens will also be strained if aliens produce revolting odors. On the other hand, consider the beneficial relations that might develop if they produced intoxicating odors via their gaseous waste products.

It is doubtful that intelligent alien beings would share any of our facial expressions and body languages, in spite of Hollywood portrayals in movies such as *Star Wars* and *Men in Black,* where alien gestures are quite easy to understand. An alien will not shake his head to mean "no," or shrink his eyes in disgust, or bare his teeth in aggression. Alien gestures and expressions will be quite exotic, making communication with them difficult. This disparity in gestures will further enhance their "alienness." When their every facial expression or gesture seems false, deceptive, or nonexistent, the aliens may seem to us more like plastic dolls or morons, even if they are highly intelligent.

Fellow scientists may smile at my speculations on the appearance of aliens, but I think most would agree that life of some form is prevalent throughout the universe. The astrophysicist Frank Drake believes that there are around 10,000 advanced extraterrestrial civilizations in our Milky Way Galaxy alone (Figure 1.13). Although, my personal belief is that this estimate is too high, for reasons I discuss in this book, the

1.13 The Milky Way galaxy. Here an astronomer looks at the stellar system of the Milky Way galaxy reduced 100,000,000,000,000,000,000 times. The head of the astronomer is approximately in the position occupied by our sun. The Milky Way includes about 40,000,000,000,000 stars distributed within a lens-shaped area about 100,000 light-years in diameter. (From George Gamow's One, Two, Three .... Infinity.*)*

mere idea of even a single other society that vaguely mirrors our own sends shivers up my spine. I agree with Frank Drake when he says:

> There is probably no quicker route to wisdom than to be the student of more-advanced civilizations. . . . Just learning of the existence of other civilizations in space—even if they are no more advanced than our own—could catapult nations into a new unity of purpose. Indeed, the search activity itself reminds us that the differences among nations are as nothing compared to the differences among worlds.

ALiEN SENSES

Forty years as an astronomer have not quelled my enthusiasm for lying outside after dark, staring up at the stars. It isn't only the beauty of the night sky that thrills me. It's the sense I have that some of those points of light are the home stars of beings not so different from us, daily cares and all, who look across space with wonder, just as we do.

—Frank Drake, *Is Anyone Out There?*

Perhaps the most important fact about the times we live in is that they are going to be different soon. We live in a world of change—what Isaac Asimov has called "a science-fiction world"—and anyone who wants to read a "realistic" fiction turns naturally to science fiction, the literature of change.

— James E. Gunn, *The New Encyclopedia of Science Fiction*

Aliens could exhibit colors that are not in our spectrum. Aliens could look like coke bottles, or could be composed of magnetic forces and undetectable to the human eye. They could be just a strange smell in the air. They could be completely digital and look like beautiful strange fractal chaotic swirls. Who knows?

—Cameron Mckechnie

In the TV series *Third Rock from the Sun,* aliens have come to Earth and used their advanced technology to install themselves in fully functional human bodies. The aliens are completely unprepared for the physical and emotional sensations of their new human bodies. Therefore, although they are intellectually superior to humans, they can appear quite foolish while experiencing many feelings for the first time. The TV series allows us to watch the aliens discover Earthlings and their quirks.

The aliens in *Third Rock* originally planned to limit their stay on Earth to a two-week fact-finding mission, but they extend the journey because they fall in love with their new bodies, feelings, and sensations. Evidently their original senses are quite different from our own. This makes me wonder: Just *how* different are alien senses from our own?

Whatever senses they have, aliens must have evolved them through time in response to their environment in order to survive. Aliens will probably need to communicate for the purposes of sexual reproduction, caring of the young, pursuing prey, escaping from danger, or—in collective arrangements—to defend, attack, and search for food in packs, flocks, colonies, or other societies. As on Earth, communication may take place in many ways, including touch, scent, sound, gestures, electrical pulses, or by other means for which we have no name. Although we humans use our visual senses to gain most information about our world, many Earthly animals have other dominant senses. Bats rely almost entirely on sonar; they hear the reflection of their high-pitched sounds from their surroundings. *Gymnarchus* fishes are those that use weak electrical discharges to sense the location of other fish. Members of the *Mormyridae* and *Gymnarchidae* fish families have electrical organs that allow the fish to distinguish prey, predators, members of their species, and obstacles in the water. Some animals appear to be sensitive to magnetism and radio waves. Rattlesnakes have infrared detectors that give them "heat pictures" of their surroundings. Alligators have tactile organs on their lower jaws with which they sense the presence of edible fish in the mud at the bottoms of rivers.

Given the diversity and range of senses on Earth, we cannot precisely predict the nature of alien senses. However, we can make educated guesses on the basis of sensory evolution on Earth. Let's start with the sense of smell.

Smelling on a Cloudy World

Humans have an amazing capacity to use their sense of smell to detect numerous chemicals. For example, researchers have identified 70 perceivable odorants coming from white bread, including alcohols, organic acids, esters, aldehydes, and ketones. One hundred three separable volatile compounds have been isolated from coffee, and at least 150 substances contribute to the flavor of coffee. Since many of these odorants are present in extremely minute quantities, the capabilities of the human olfactory epithelium, usually regarded as being lame compared to other mammals', seem remarkable. For substances called mercaptans (they are present in the skunk odorant), only about 40 receptor cells in the human nose need be stimulated by less than nine molecules each to give a detectable odor sensation. Still, when it comes to odor detection we are relative cripples, compared to other Earthly animals—and possibly also when compared to alien life-forms.

Imagine a race of aliens that develops on a dimly lit world perpetually shrouded in clouds so that vision would be less useful for survival than on Earth. What would life be like for them? These aliens may have a particularly sensitive sense of smell. In the same way that we *see* faces, trees, and mountains—and form an impression of our world primarily through our visual sense—these creatures would "see" the world primarily through odor. Characters on a printed page, or pictures in magazines, might be represented by different long-lasting scents positioned at different locations on a page. These aliens would recognize friends and relatives by odor. Although they could not visually perceive the sun through their heavy clouds, they might smell the location of the sun by various localized chemical reactions triggered by the sunlight impinging on molecules in the atmosphere. For example, when ultraviolet rays interact with oxygen they form ozone, which may be detected by these creatures. Other kinds of electromagnetic radiation, such as X-rays and radioactivity, effect subtly the materials on which they impinge and might also be detectable by odor.

On Earth, the animal with the most acute sense of smell is the male emperor moth (*Eudia pavonia*) which can detect the sex attractant of the virgin female at the almost unbelievable range of 6.8 miles (11 km) upwind. If you had this ability, you could smell someone standing at the top of Mount Everest while you stood all the way at the bottom. The moth's scent is a form of alcohol of which the female carries less than 0.0000015 grain (0.0001 mg).

Other moths also exhibit impressive olfactory feats. For example, the antenna stem of a silkworm moth contains roughly 35,000 olfactory hairs, which suggests to some scientists that their sense of smell ranks on a par with vision in many animals, both in sensitivity and in discrimination. The chemoreception on the male moth's antennae is so sensitive that they can detect a single molecule of scent.

Ever wonder how mosquitoes can find you so easily? Each time you breathe out carbon dioxide, you're telling mosquitoes that there's a bloody vertebrate nearby. Mosquitoes have carbon dioxide receptors on little feelers called palpi and can detect a plume of gas from about 50 feet (15 m) away. They also can sense lactic acid, a volatile chemical exuded from human skin. If the aliens from the cloudy world had the olfactory abilities of mosquitoes, and they visited Earth with evil intentions, it would be hard for us to hide!

Snakes have the chemical equivalent of stereoscopic vision, in the form of odor receptors on their tongues. By responding to the relative number of odors on either side of their tongue, snakes can pinpoint prey, enemies, or mates.

Let's call the race of aliens with the supersensitive sense of smell Smellers. They would have long noses or proboscises because long appendages help them sample particular directions and objects, thus giving them more control and understanding of their surroundings. Their primary sense being smell, in some ways they are better off than we are with our primary vision sense. For one thing, Smellers can know precisely how long ago you were sitting in a chair and the direction you went after leaving the chair. Olfactory communication gives information about the past and present. Once the Smellers had familiarized themselves with human structure and psychological responses, they could determine gender, health, and even moods using their sense of smell. Smellers could even use their olfactory sense to see around corners in darkness. In the same way we can sense depth using two eyes (binocular vision) an alien could sense depth using odor. Their ability to sense both past and present would affect their entire way of thinking. For example, their demarcation of hours in a day may be fuzzy—owing both to their analog odor sense and to the fact that their cloud-covered world does not clearly demarcate sundown and sunset.

Since Smellers can sense the gradual diffusion of odorants, their number system might use numbers representing gradients between each integer. The number 1 would represent a field extending from 1 to 2, and so on. As a result, Smellers' mathematical calculations would be expressed in symbols of probability and would utilize the concept of fuzzy logic. (On Earth fuzzy logic was first described in the 1960s, and it deals with probabilities or degrees of truth on a continuum of values ranging from 0 to 1.) Perhaps as Smeller technology advanced, the Smellers would find a way to traduce their olfactory signals into magnetic signals that are more easily stored, manipulated, or transmitted via radio to remote locations.

We would probably have a difficult time negotiating with the alien Smellers, because emotion is mirrored by an animal's scent, at least on Earth. Our moods and intentions might be broadcast before we even said a word. On their own world, the Smellers' ability to sense moods might lead to a more harmonious living and family structure. Before a teenager was about to burst out screaming at his parents both might better understand, or at least be prepared for, any escalating argument. On the other hand, devious Smellers might learn to hide their emotion-induced odors, much as certain criminals are able to outsmart lie detector tests. Doris and David Jonas, in their remarkable book *Other Senses, Other Worlds,* suggest that aliens relying primarily on smell might carry perfume packets in their pockets which would add to their personal identifying odors and provide information about their rank or status in society.

If you were to visit the world of the Smellers, the architecture would seem visually quite boring. Instead of paintings hanging on the walls of their home, they might use certain aromatic woods and other odor-producing compounds strategically positioned on their walls. Their counterparts of Picasso and Rembrandt wouldn't make paintings but would create exquisite compositions of bold and subtle perfumes. Smeller equivalents of *Playboy* magazine would be visually meaningless but awash in erotic aromas. Their culinary arts would be like our visual or auditory arts: Eating a meal with all its special flavors would be akin to listening to Beethoven's Fifth Symphony. If the primary sense of all the animals on their world was the sense of smell, there would be no colorful flowers, peacock's tails, or beautiful butterflies. Their world might look gray and drab. . . . But instead of visual beauty, an enchanting panoply of odors would lure insects to flowers, birds to their nests, and aliens to their lovers.

How would we respond to a visit from the Smellers? The answer depends to a large part on our acceptance of behaviors different from our own. Just imagine how strange it would be to encounter an intelligent alien that marks its trails with urine, just as dogs do on Earth. We could ask similar questions about aliens with other acute senses. For example, what would it be like to interact with an alien possessing a sense of hearing as acute as that of our Earthly foxes and wolves, which can hear our wristwatches ticking 30 feet (9 m) away or can hear insects moving? Aliens with sufficiently powerful senses would appear Godlike to us. The greater their senses, the closer they come to seeming omniscient.

Alien Eyes

When I think about the possibility of traveling to alien worlds, I always remind myself that the most exotic journey would not be to see a thousand different worlds, but to see a single world through the eyes of a thousand different aliens. I mean this not only in the symbolic sense of viewing the world from various alien mental perspectives, but also literally: seeing through *eyes* sensitive to strange, nonvisible parts of the electromagnetic spectrum, seeing in all directions simultaneously, or seeing events that are so quick that they are a mere blur to the human eye. Aliens on worlds illuminated by a sun would have vision because of its survival value. On Earth, eyes of various kinds have evolved numerous times in different animal groups. Even certain one-celled organisms have eyelike structures called eyespots.

Earthly eyes fall broadly into two categories: direction (non-image-forming) and image-forming. Direction eyes are found in many worms, mollusks, cnidarians, echinoderms, and other invertebrates. These eyes have light-sensitive cells and often a cup-shaped shield that nearly surrounds the sensor. In bilaterally symmetrical animals the eyes are usually paired. Image-forming eyes are found in certain mollusks (cephalopods and some bivalves), most arthropods, and nearly all vertebrates.

By studying the creatures of the Earth, we can hypothesize on the diversity of alien eyes and visual perceptions. Aliens would no doubt see a world different from the one that we do. To best understand this, consider the Indian luna moth (Figure 2.1), which has a wingspread of about 4 inches (10 cm). To our eyes, both the male and female moths are light green and indistinguishable from each other. But the luna moths themselves perceive in the ultraviolet range of light,[1] and to them the female

*2.1 Luna moth. To our eyes, both the male
and female moths are light green and indistinguishable from each other. But the luna
moths themselves perceive in the ultraviolet
range of light, and to them the female
looks quite different from the male.*

looks quite different from the male. Other creatures have a hard time seeing the moths when they rest on green leaves, but luna moths are not camouflaged to one another since they see each other as brilliantly colored.

On Earth, bees' sight also extends into the ultraviolet range, although they do not see as far as we do into the red range of the spectrum. When we look at a violet flower, we do not see the same thing
that bees see. In fact, many flowers have beautiful patterns that only
bees can see to guide them to the flower. These attractive and intricate
patterns are totally hidden to human perception.

Bees also see the world differently from us because of their remarkable flicker-fusion rate. "Flicker-fusion" refers to the number of frames
per second at which sequential images are no longer seen as separate.
Humans can distinguish from 16 to 24 flickers per second. Motion pictures show about 30 still frames per second. If an alien had the flicker-

fusion of a bee, the alien could see 265 separate flickers per second before fusion takes place. Our movies would seem like a slide lecture to them. As a result of their higher flicker-fusion, bees can see objects moving at speeds higher than those we can see—to us the image becomes a blur or eventually invisible.

Imagine aliens who could make gestures so quickly that we couldn't see them, but the aliens would have no trouble seeing them. Imagine what it would be like for us to see the vibration of a fly's wing in flight or the intricate array of droplets splashed when a raindrop hits a puddle.

If we were able to extend our current senses in range and intensity, we could glimpse alien sense domains. For example, if we possessed sharper sight we would see things that are too small, too fast, too dim, or too transparent for us to see now. We can get an inkling of such perceptions

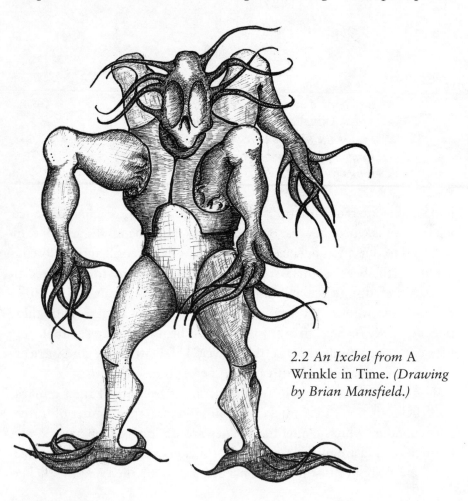

2.2 An Ixchel from A Wrinkle in Time. (Drawing by Brian Mansfield.)

using special cameras, computer-enhanced images, night-vision goggles, slow-motion photography, and panoramic lenses, but if we had grown up from birth with these visual skills, our species would be transformed into something quite unusual. Our art would change, our perception of human beauty would change, our ability to diagnose diseases would change, and even our religions would change. If only a handful of people had these abilities, would they be hailed as gods or messiahs?

Although most aliens in the science fiction literature are described as having eyes, there are notable exceptions. My favorite example is the Ixchel in Madeleine L'Engle's *A Wrinkle in Time*. The Ixchel (Figures 2.2 and 2.3) have graceful tentacles extending from each of their four powerful arms; these tentacles act both as fingers and as speech organs.

2.3 Another artist's vision of an Ixchel (Figure 2.2). (Drawing by Ken DeVries.)

Softly waving tentacles on the head function as receptors of sound and thought. Since the Ixchel come from a planet in which the atmosphere is opaque, the Ixchel never developed eyes and cannot understand what humans describe as sight.

Finally, I like to speculate on the facial characteristics of aliens and humans living in a hypothetical universe where light totally penetrated objects. Usually, some light rays are reflected back into our eyes, thereby allowing us to see objects. If light didn't bounce off matter, we would not have eyes or a visual cortex of the brain. What would aliens be like that could "see" gravitational pull? Since gravity does not reflect at all, it's hard to imagine what creatures with this ability might sense. As writer Joel Achenbach has pointed out, if we learned to see the universe through the "gravitational wave band," there's no telling what we would find. For example, when radio astronomy was invented after World War II, we discovered that the seemingly tranquil universe (as seen in the visual band) was actually awash in violent quasars, pulsars, and cacophonous radio signals that defy understanding.

Electrical Senses

Let's turn our attention from the visual sense to the electrical sense exhibited by several animals on Earth. Any aquatic aliens may have a keen electrical sense. Like the echolocation of bats, which registers the reverberations of high-frequency sound waves, "electric location" may exist in aliens; impulses may also be emitted by aliens to "sound" nearby objects. Alternatively, an alien may surround itself with an electric field so that any object penetrating the field distorts it and becomes apparent to the alien at its center. The first form of electrical sense would be used when needed, but the second would be maintained as a constant state—like a surrounding sensitive web. Many objects that were transparent to an alien with an electrical sense may be opaque for us, and vice versa.

On Earth, true electric organs have evolved independently in at least six different groups of fishes, indicating that there is a tendency for this kind of sense to evolve on Earthlike planets. Examples of fishes with the electrical sense are the skate, knife fish, elephant fish, electric eel, electric catfish, electric ray, and the stargazer, an elongated fish that buries itself at the bottom of the ocean.

The German zoologist Henning Scheich discovered that the fishes of the genus *Eigenmannia* have elaborate social behaviors based on electricity. For example, when two of the creatures meet, they have the ability to adjust their frequencies up or down to avoid jamming each other's signals. Mormyrid fishes (freshwater African fishes found in muddy waters) have specialized lateral organs that emit an irregular discharge, which becomes regular as soon as another fish approaches. The fish can also turn off its electrical field so that it is electrically undetectable as it hides or listens, or it can raise its frequency to assert dominance when other mormyrid fish intrude into its territory.

Fishes of the genus *Sternopygus* use electricity for love. For example, if a female fish swims past an adult male of her own species, she "turns him on": His steady-state single-frequency discharge changes into a chaotic electric lovesong. These fishes also can use the electricity for recognizing different individuals of the species.

An alien with an electrical sense would perceive a world fundamentally different from the world we perceive. Shape, form, transparency—all terms become blurred. If aliens could sense radio waves, and thereby evolve to transmit them, they might not need any advanced technology to send signals over long distances. Their world might be without light, or they could live in a dark cave, but they still could perceive the world around them.

What would intelligent, technologically advanced, electric aliens look like? Since they rely on the electrical characteristics of their bodies, they may wear few clothes. As a result, we might expect their skin to be tough or scaly like a fish's. Might these aliens have shells or exoskeletons like a lobster? Because we have not discovered any very intelligent Earthly animals with external shells, it's likely that a highly intelligent electric alien does not possess a hard shell, which would restrict the alien's range of motion and manual dexterity. Technological aliens with hard shells are probably even less likely. (Imagine your dentist working on you with lobster claws . . .)

The electric alien could sense our appearances, but because we emit no strong fields, it might suspect we were hiding something, just as we might be suspicious of an alien that was visually camouflaged and hard for us to see. If their electric senses permitted only short-range observation, their gods may not be based on the heavens or the stars but rather on water, lava, and rock.

The Senses of Immortals

The Old Ones in H. P. Lovecraft's *At the Mountains of Madness* are incredibly tough and durable creatures, having characteristics of both plants and animals (Figures 2.4 and 2.5). They also possess an extraordinary array of senses to help them survive. Hairlike projections and eyes on stalks at the top of their heads permit vision. The colorful, prismatic hairs seem to supplement the vision of the eyes, and in the absence of visible light, the species is able to "see" using the hairs. Their

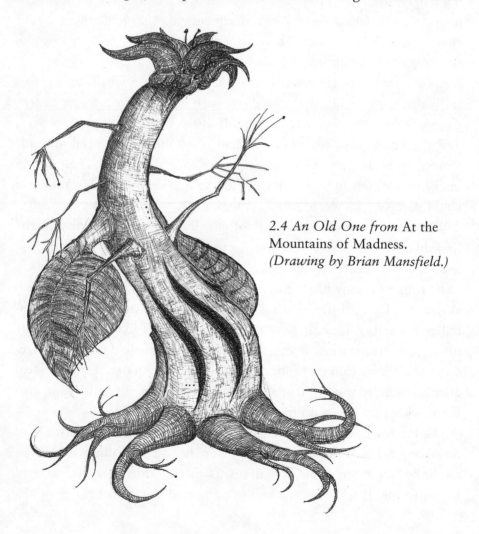

2.4 An Old One from At the Mountains of Madness. *(Drawing by Brian Mansfield.)*

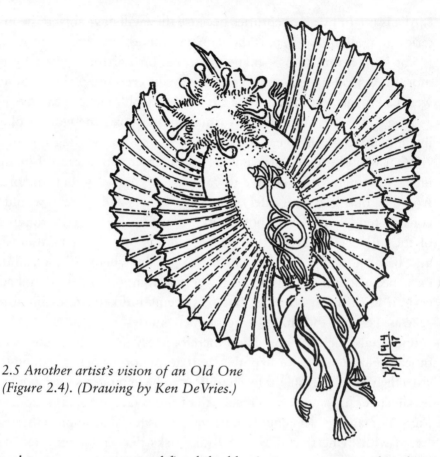

*2.5 Another artist's vision of an Old One
(Figure 2.4). (Drawing by Ken DeVries.)*

complex nervous system and five-lobed brain process senses other than
the human ones of sight, smell, hearing, touch, and taste. When the
Old Ones open their eyes and fully retract their eyelids, virtually the
entire surface of the eye is apparent.

Old Ones can survive for long times in space, storing air, food, and
minerals and sailing the solar winds on large membranous wings. Millions of years ago the Old Ones came to Earth, where they lived under
the sea, building cities and exploring their new world. Later they mi-
grated to land.

As I gaze upon images of the Old Ones, I enjoy wondering about the
typical lifespans of aliens. The SETI pioneer Frank Drake believes that
any aliens we encounter will be immortal. In 1976 he wrote, "It has
been said that when we first discover other civilizations in space, we
will be the dumbest of them all. This is true, but more that than, we
will probably be the only mortal civilization."

Aliens might live for centuries because they will have solved the mysteries of aging or can repair the damage caused by aging. Immortality is not such a rare thing—many creatures on Earth are virtually immortal. As just one example, consider desert creosote plants in Southwest California, estimated to be 11,700 years old. Lichens can live just as long.[2] In 1997, scientists in Tasmania discovered the world's oldest living plant, a 43,000-year-old *Lomatia tasmanica*, or King's holly.

What would it be like for humans to interact with a race of immortals? How might they behave toward us? My guess is that immortals would be obsessed with safety, and their devices and vehicles would be constructed to present no possibility of fatality. Perhaps the roadways of the immortals have very low speed limits to prevent fatal crashes and because they're not in a hurry to get anywhere. Wars wouldn't exist because immortals would never risk fighting, unless they had a certain faith in an afterlife, but this seems unlikely in an immortal species. Perhaps they might employ mercenaries . . .

Immortals might be bored and therefore very interested in contacting other civilizations. If we hypothesize that immortals fear anything that risks their lives, they might teach us how to be immortal so we would be less likely to have wars with them. All of this is creative speculation, and I'm sure many readers could argue the opposite—that immortals are so bored with living that they easily take risks that endanger their lives. Does life become more precious or less precious when a species lives for an eternity? How would overpopulation problems be solved?

Communication

So far we've been focusing on alien senses, but through what mechanism would aliens communicate? If you found a dead alien in your backyard, you could guess at its possible forms of communication by examining its sense organs. Every one of the ways in which a creature senses the world suggests a possible way it can communicate. For example, an alien with eyes that are highly sensitive to infrared radiation might communicate by altering heat distribution patterns on its body. Sign language on Earth suggests that communicating through gestures can be efficient. Perhaps even direct brain-to-brain communication is possible through temporary or permanent nerve interconnections. If an alien race evolved brain-to-brain communication among hundreds of individuals, an entire group could take collective actions to produce artworks, music, religions, and technologies we cannot imagine.

Would aliens talk? We've already discussed the Ixchel, who use their arm tentacles to produce musical speech, and the Cygnans, who use their spiked, tubular tongues to produce a musical language based on chords. Certainly, humanlike vocal communication is not a prerequisite for communication, despite the fact that Hollywood movies require humanlike vocalizations in many aliens so that actors can easily convey their intentions. In fact, we're lucky we can talk at all. Humans can produce complex utterances only because we went from a four-legged to a two-legged, upright posture, which rearranged the structure of our vocal organs in relation to each other. Many evolutionary biologists feel that without this change of posture, there would not have been the change in mental development that led to the complexities of human speech. This is one reason why monkeys can't be trained to talk. Even though chimpanzees are anatomically extremely close to humans, their vocal organs, such as the larynx and palate, are positioned in a way that makes it impossible to produce the array of sounds we can make. Perhaps an alien with a more versatile way of producing sounds would consider the human vocal apparatus to be similarly crude.

Despite the various wild speculations in this chapter, we can be sure that alien senses, whatever they may be, will permit aliens to experience a universe entirely different from the one we experience. They will see things we don't see, understand relationships we can never understand, and think thoughts we can never think. The problem of sharing the experiences of aliens that rely on different senses is a profound one. Humans can barely imagine an alien's *Umwelt*—a German word used by animal behaviorists to denote the environment as perceived by an animal. The closer an alien is to us from a sensory standpoint, the easier it is to enter its *Umwelt*. However, our sensorium (that part of the mind where our senses may come together) is just one form of sensorium. All about us, parted from us by thin veils, there lie potential forms of consciousness that are entirely different. Our descriptions of the universe are not comprehensive if they do not consider the possibility of alien perceptions and consciousnesses. Our speculations and studies determine attitudes though they cannot furnish formulas, and open wide vistas though they provide no map.

How would humanity's perception of the world change if we could learn from aliens? If the properties we assign to the natural world are expressions of the way we think and our capacity for understanding, then encounters with aliens will *change* those properties. The universe made visible by an alien encounter seems limitless.

LIFE At THE EDGE

Out beyond our world there are, elsewhere, other assemblages of matter making other worlds. Ours is not the only one in air's embrace.

—Lucretius, Roman philosopher
of the first century B.C.

One of the reasons I'm not a big fan of science fiction is that none of the aliens are nearly as weird as my own invertebrates.

—Janet Leonard, Marine Science Center,
Oregon State University

I Have No Mouth but I Must Scream

Science fiction writers have long imagined aliens living in extreme conditions: in solar atmospheres, in the cores of planets, on the surface of high-gravity neutron stars, in the vacuum of space, in iron-vapor atmospheres, and even in the industrial corridors of New Jersey. Larry Niven's *Flatlander* and *Known Space* aliens had cells based on superconductive helium instead of water. In David Brin's *Sundiver*, aliens lived in the upper layers of the Sun. In Arthur C. Clarke's *Childhood's End*, there were alien crystal formations on planets with wildly varying temperatures. In James White's *Sector General* series, aliens thrived on hard radiation.

How far-fetched are these scenarios? Could life exist under the extreme conditions on other planets in our solar systems?

As we wonder about alien life, it is useful to look at our own world to learn about conditions that may support life elsewhere. Although the diverse Earthly habitats are not microcosms for planetary environments throughout the universe, in the last few years it has become clear that Earthly life can survive, and even thrive, under environmental conditions that seem quite alien.

The masters of life in extreme environments are microscopic organisms such as bacteria and fungi, although larger creatures (for example, insects and crustaceans) can survive under surprisingly nasty conditions. Because terrestrial life is water-based, many adaptations to harsh conditions center around desiccation, resistance to freezing, or preventing proteins from melting at high temperatures. Some organisms can thrive under conditions that are highly toxic to most other forms of life, such as the presence of heavy metals. As just one example, in 1997 Stephen Zinder of Cornell University discovered bacteria that flourish on perchloroethylene (PCE) and trichloroethylene (TCE). Such solvents are normally used to clean clothes, machines, and electronic parts.

Earthly life-forms that flourish in extreme environmental conditions are known as *extremophiles*. Of course, the term "extremophile" reflects a bias: Aliens living in environmental extremes would think *we* were the extremophiles, because from their point of view, we are the ones who live under extreme conditions. For example, life-forms living in the vacuum of space, such as the Starseeds and Outsiders from Larry Niven's *Known Space*, consider the surface of Earth—with its "crushing" gravity, dense "corrosive" atmosphere, and lack of radiation—to be extreme. Given this caveat, I think that the weirdest Earthly extremophiles are those living near hot, deep-sea, hydrothermal vents—fissures in the sea floor from which pours hot seawater filled with sulfides and metals. Organisms in this unique environment exist in the complete absence of light, adjacent to magma-heated plumes of fluid in excess of 752 degrees F (400 degrees C). Bacteria in these environments have been cultured at temperatures of 248 degrees F (120 degrees C), and the maximum growth temperature may be higher than 302 degrees F (150 degrees C). As a comparison, water boils at 212 degrees F (100 degrees C) and paper catches fire around 415 degrees F (213 degrees C).

Let's begin our discussion of extremophiles with invertebrates, animals that lack backbones. More than 90 percent of living animals are invertebrates. Worldwide in distribution, they range in size from minute protozoans to giant squids. Of the 22 invertebrate groups known as *phyla*, at least 9 are represented at the vents. Some of these animals tend to live in the cooler waters surrounding the vents, where they must tolerate extremely high concentrations of heavy metals and sulfur-containing compounds. While most Earthly animals are ultimately dependent on light energy harvested by plants, vent animals depend on bacterial chemical reactions involving sulfide. For example, large worms in the ocean trenches, living where it's too deep for sunlight to penetrate, use the energy supplied by deadly mixtures of sulfuric gasses ("deadly" to other life-forms, that is). These 10-foot-long (3 m) worms under 8,000 feet (2,400 m) of water are stranger than anything out of a Hollywood science fiction movie. The giant worms have no mouths. They can't eat.

The worms' digestive systems are oozing with reddish bacteria that acquire energy for the worms by digesting hydrogen sulfide from the vents. In this acidic environment, the worms grow 0.08 inch (2 mm) a day, making them the fastest-growing marine invertebrates on the planet. It's hard to believe creatures such as these exist on Earth. In fact, when I have to design aliens for my science fiction novels, I get ideas from photographs of Earthly invertebrates. There's nothing stranger on our planet.

In addition to the mouthless vent worms, there are also giant clams, red from the presence of similar bacteria, thriving in water as hot as 680 degrees F (360 degrees C), much hotter than the temperature at which paper bursts into flame.

Let's discuss the bacteria that thrive in the vent worms' guts. Traditional microorganisms are killed by heating to 212 degrees F (100 degrees C). However, extremely *thermophilic* (heat-loving) bacteria, or *hyperthermophiles* (extreme heat lovers), not only survive exposure to such temperatures but also grow optimally above 212 degrees F (100 degrees C)—temperatures at which the water remains liquid only because of the extremely high pressures. The bacteria go by such imaginative names as *Pyrococcus furiosus* or *Methanothermus fervidus* and are found in naturally hot environments and in human-made environments such as hot-water tanks. In contrast to hyperthermophilic bacteria, less extreme, thermophilic bacteria grow at temperatures higher than 140 degrees F (60 degrees C). These temperatures are encountered in rotting compost piles, hot springs, and also in oceanic geothermal vents.

Hydrothermal vents could have supported the first forms of life, and the deep-sea environment may have protected them from cataclysmic meteorite impacts occurring on ancient Earth. For example, primitive bacteria called *archaebacteria* have evolved little since the time of their origin billions of years ago, and they prefer hot temperatures as high as 248 degrees F (120 degrees C). The early Earth contained no molecular oxygen, and since the hyperthermophiles live in the absence of free oxygen, they would have been quite happy billions of years ago. Scientists are researching how the organisms maintain the structural integrity of their components, particularly since proteins and genetic materials (DNA and RNA) are normally quite heat-sensitive.

The potential for commercial uses of the high-temperature stability, or thermostability, of the enzymes produced by extremophiles found in hot springs and alkaline lakes has attracted the interest of genetic-engineering companies. In 1997, Genencor International introduced a new detergent additive to make cotton clothes look like new through hundreds of washings. The additive, an enzyme called cellulase 103, was taken from an extremophile. It works at the pH of soapy wash water—hot or cold (pH expresses the relative acidity or alkalinity of a solution). The bacteria are collected from soda lakes (very alkaline bodies of water) on several continents—although Genencor won't reveal the supersecret location!

Aliens Down Under

Perhaps the most likely place to find alien life is *beneath* the ground on planets or moons. Incredibly large numbers of life-forms live miles under the Earth's surface without any "help" from the surface in the form of light, air, or nutrients. As we discussed with the hyperthermophilic vent creatures, if bacteria (and archaebacteria) can survive in these seemingly strange environments, then it's also possible that life could have started there. In many ways, the surface of the Earth has been particularly uncomfortable for much of Earth's history. The surface has often been bombarded by giant meteorites and high doses of ultraviolet rays from the Sun. There have been huge volcanic eruptions, thick and deadly gases from the interior, and solar instabilities that would have made life a living hell.[1] However, deep rocks would have provided protection from the sterilizing temperatures and radiation.

Today, it's possible that creatures are living comfortably beneath the dusty red surface of Mars and other planetary bodies. Although the Martian surface appears to be inhospitable because it lacks liquid water, fluids may flow through the warmer interior of the planet. Mars used to contain lots of water, as evidenced by networks of channels like those formed on Earth by drainage of rainy areas.

Amazingly, most geomicrobiologists—biologists who study, among other things, the physical and chemical interactions of microorganisms with Earth—believe that there could be as much life hidden below ground on Earth as there is above. The astrophysicist Thomas Gold, of Cornell University, a steadfast proponent of deep life's ubiquity, has calculated that the weight of all subterranean microbes could equal that of all organisms above the surface.

Just what is the evidence for alien life living inside the bowels of planets? The best support comes from "deep biology" studies on Earth. Deep biology includes the study of subsurface bacteria and archaea, microscopic organisms that include the hyperthermophiles. For a moment, imagine yourself walking in a deep, underground mine with the Princeton University geologist Tullis C. Onstott. Want a peek at hell? With your shirts drenched from the 100 percent humidity, you make your way through the deepest gold mine in South Africa, where the temperature of the rock reaches 1,400 degrees F (600 degrees C) and there's not the slightest trace of sunlight. After an hour of hiking through passages 2 miles (3.5 km) below Earth's surface, you reach a recently blasted section of tunnel and take a hammer to the wall knocking loose nuggets of rock that have spent the last 3 billion years locked underground. Tests on the rocks reveal that thermophilic bacteria somehow manage to survive, even at the extreme depths. This scenario is not fiction, and there are numerous bacteria living in the hot bowels of Earth.

A decade ago, the idea of life flourishing so far beneath Earth's surface would have seemed unlikely. In fact, I was taught that life inhabited only a thin skin of territory at Earth's surface. I learned that the oceans, the air, the ground, and even the soil teem with animals, plants, and microorganisms, but my training in biology never prepared me for the idea that animals could flourish locked within hot rocks deep underground and surrounded by toxic chemicals.

Today we know better. In the late 1980s, researchers found microbes living in rock 1,640 feet (500 m) below the surface in South Carolina. In the last four years, Onstott and other researchers have pushed the envelope of life much deeper, extending it to around 2 miles (3.5 km) below ground. In many cases, the pore sizes in the rock samples are so small that they wouldn't even allow outside bacteria to penetrate. These microbes are living fossils, prisoners of the deep for millions of years. As a result of the scarcity of nutrients, life at great depths is extremely slow paced. The cells may divide once a year or once a century, rather than every few minutes, as in an infection, or every few hours, as in soil. The line between life and death becomes indistinct and has no meaning whatsoever.

Compared to some of the dense rocks deep within Earth, porous rocks near the surface are a relative heaven for the rock dweller because these rocks can store water in their pores and admit sunlight, allowing photosynthesis. These rocks are relatively comfortable even in hot environments because they filter the extremes of strong light that can fry microorganisms in the desert. In fact, in deserts, more photosynthesis occurs inside rocks than in the soil. On Mars, life that could have been plentiful on the surface may have retreated inside rocks when the Martian surface lost its rivers.

On Earth, life is also in rocks beneath the ocean floor, for example, below the 37,000-mile-long (60,000 km) system of volcanic mid-ocean ridges that girdles the planet. Researchers have even coaxed a variety of microbes to grow on a diet of crushed basaltic rock that reacts with oxygen-free water to produce hydrogen—the only energy source in their ecosystem. If Mars ever hosted life, it may have permeated the subsurfaces, which may have been as hospitable to life as is Earth's subsurface: mild temperatures, liquid water, dissolved minerals, and plenty of rock surface.

Attack of the Metal-Eating Aliens

If you woke up one morning to hear the perky host of *The Today Show* telling you that all our industries, appliances, and computers were being attacked by metal-eating aliens, and that these creatures couldn't be killed by doses of radiation millions of times greater than would kill a human, you'd be pretty scared, right? Yet such creatures, called *metalophiles,* are already quite common right here on Earth and provide models for possible life-forms on other worlds. These creatures are archaea, primitive bacterialike organisms that we discuss in greater detail in this section.

In 1996, J. Craig Venter's team at the Institute for Genomic Research in Rockville, Maryland, learned more about a microbe that can live at freezing temperatures, eat metal, and tolerate huge doses of radiation with apparent gusto. After sequencing its complete genetic structure, which includes 1,738 genes, they discovered that two thirds of its genes don't look like anything scientists have ever seen before. Some of the genes are similar to those of humans while others resemble those of bacteria.

A member of the plentiful but perplexing group of one-celled organisms called archaea, the microbe has genes that prove the creature belongs to a third branch of life, totally different from Earth's two other known branches: bacteria and blue-green algae on one branch, and all the members of the plant and animal kingdoms on the other. Believe it or not, *most* Earthly life is one-celled, and all eukarya, or multicellular organisms (from plants to ants to people), are a little twig protruding from a humongous microbial tree. Needless to say, this lopsided view of biology was not quickly accepted by biologists when it was first proposed.

In the group archaea there may be millions of different organisms. They make up half of all the living mass on Earth and include microbes that can withstand radiation in doses rated at 2 million rads—450 rads are fatal to any human. Other members of the class thrive at temperatures far below freezing or can withstand temperatures hotter than boiling water, or live only by consuming metals and minerals such as sulfur.

The study of archaea accelerated in 1982 when oceanographers aboard a submarine 1,000 miles (1,613 km) off the coast of Baja California found an undersea volcanic vent called a white smoker spewing out a dense white cloud in bursts of hot water. The submarine crew collected samples of the hot water and found it contained a strange and unknown organism.

Keeping such creatures alive in the laboratory is quite difficult. In 1996, Craig Venter's group cultured them at high temperatures inside pressurized containers that were vented to prevent the methane the organisms produced from exploding. The microbe called *Methanococcus jannaschii* normally lives in vents 2 miles (3.2 km) under the Pacific, where water pressure is hundreds of times greater than at sea level. It thrives in total darkness at lethal temperatures of 185 degrees F (85 degrees C), and oxygen kills it, like many bacteria, instantly. It subsists entirely on carbon dioxide, hydrogen, and nitrogen.

The gene sequence of one of these creatures suggests that it shares a common evolutionary ancestor with bacteria. Because the earliest indications of life on Earth have been 3.6-billion-year-old fossils of bacteria from western Australia and 3.85-billion-year-old sediments from eastern Greenland, the unknown ancestors of the archaea must have been alive even earlier—perhaps as far back as 4 billion years ago, when Earth and the Solar System's planets were very young. Some scientists suggest that hypothetical fossils found on Mars could be one-celled archaea.

In the future, the microbe will be increasingly useful to humanity, because it emits large amounts of methane, an explosive natural gas widely used in industry and an energy source. The varied methane-producing organisms in the archaea group emit at least 200 million tons of methane a year. These metal eaters may also be useful in yielding new pharmaceutical products and in cleaning hazardous-waste sites, for example those containing toxic heavy metals.

Eukaryotes

Traditionally, earthly life has been divided into two basic categories: life-forms whose cells have no nucleus (*prokaryotes*) and those whose genetic material (usually DNA) is concentrated in a nucleus (*eukaryotes*), the later including all higher plants and animals. As alluded to in the last section, more recently some biologists have divided the most primitive prokaryotes into two categories: the archaebacteria and the eubacteria, also referred to more simply as archaea and bacteria.

So far we've been discussing rather primitive life-forms, bacteria and archaea, living in extreme environments of temperature, light, and chemistry. Are there a significant number of extremophile *eukaryotes*?[2]

In this section we extend our search to include more complex extremophiles in a range of categories. Get set to meet the scotophiles, anaerobes, thermophiles, psychrophiles, acidophiles, alkalophiles, halophiles, and barophiles. The emphasis of this section is the same as in previous ones: No matter where we look on Earth—under conditions of extreme temperature cold, acidity, and pressure—there is always an abundance of life.

Scotophiles—Dark Lovers

You are with Captain Picard and Lieutenant Worf from *Star Trek: The Next Generation* exploring a subterranean air chamber on a new planet. The pocket has never come in contact with sunlight, so you expect to find few life-forms larger than bacteria. Yet, as you shine a light into the chamber, you see huge aliens with jointed legs and extra-optic sensory structures—"antennae" of outrageously large sizes.

Sound far-fetched? Not at all. Such a scene was observed recently on Earth.

Even totally dark caves crawl with large organisms of every variety. Consider, for example, Movile Cave in Dobrogea, Romania, isolated from the world for millions of years, forever dark, yet oozing with life. It's a scene straight out of an *X-Files* episode, or perhaps *Arachnophobia*. Despite the fact that the cave receives no energy from the sun, a unique community of alien-looking animals stalks its tortuous inner reaches: hundreds of spiders, previously unknown microbes, water scorpions, predatory leeches, isopods, pillbugs, springtails, millipedes, bristletails, and other *troglodytes* (cave dwellers) of all kinds . . .

What would an alien *scotophile* (literally, "dark lover") look like? On Earth, these creatures often use their antennae or legs to "see" by feeling their way around. Also, because they can't be seen by predators, troglodytes never need to develop coloring for camouflage, so many are pale or white—and some are so translucent that you can see their blood flowing. If we want to search for life on Mars, we must look in a Martian version of Movile Cave where liquid water could exist and light need not penetrate.

In 1996, scientists were able to gain access to Movile Cave through an artificial entrance shaft, created by accident during a construction project. Once in the cave, they found abundant fauna in the dank chambers—some 47 animal species to date. Thirty of the 47 species were previously unknown. In a pattern called troglomorphy, all creatures show a reduction or loss of eyes and pigmentation, enlargement of appendages, and gigantic antennae. The ancestors of some of these species may have become isolated from their surface-dwelling relatives more than 5 million years ago, when the climate of southern Romania became very dry. Today, food and bacteria don't enter the cave from the surface. This means that the sole source of food for those creatures at the bottom of the food chain is hydrogen sulfide and methane from groundwater trickling through the cave's inner chambers. In a strange way, Movile Cave is a time capsule—a tiny sealed-off portion of the world. Through time its creatures have evolved into a myriad of specialized forms. If humans were placed in an environment like Movile Cave, with sufficient oxygen leaking into the cave via minute cracks, what kinds of creatures would we evolve into in the next 5 million years? Would our eyes fade away? Would our fingertips become extrasensitive? Would we become like aliens on a far-away world?

Anaerobes–Oxygen Haters

Most species of free-living, microscopic protozoans appear to be *obligate aerobes*; that is, they cannot survive without oxygen.[3] Various processes of respiration that take place in cell compartments (called the mitochondria) require oxygen. On the other hand, eukaryotic organisms that are *obligate anaerobes*, in which metabolism must take place in the absence of oxygen, are much less common. Nevertheless, we know from studies of life on Earth that rather developed organisms can exist without oxygen. We may classify these creatures as extremophiles, because to us they are living under extreme conditions. For example, many free-living ciliates (microscopic, single-celled animals with hairlike projections) can live in anaerobic habitats. These ciliates use cell compartments called hydrogenosomes instead of mitochondria to make energy. Anaerobic ciliates that require no oxygen include the yeasts, various parasitic organisms in the gastrointestinal tracts of humans, and organisms associated with sulfide-containing sediments.

Thermophiles—Heat Lovers

We've already discussed some of the primitive heat-loving bacteria and their kin. Turning to more advanced, high-temperature eukaryotes we find the acidophilic phototroph (acid-loving plant) *Cyanidium caldarium*—a "red alga" that can still grow at 134 degrees F (57 degrees C). Virtually all the hot, acid soils and waters in the world are colonized by *Cyanidium*.[4]

Over recent years, there has been much heated discussion whether it is possible for eukaryotic architecture to evolve a hyperthermophile that thrives in boiling water. Scientists are not certain that the central biochemical machinery—nucleic acid transcription and translation in eukaryotes—can operate at high temperatures. Also, the cell's membrane composition must be able to retain the required degree of fluidity for proper function. So far, all known hyperthermophiles contain reverse gyrase, an enzyme that induces positive super-coiling of DNA to enhance its thermal stability. Currently we do not yet understand all the protective mechanisms operating to allow cells like the archaean *Pyrococcus* to thrive in temperatures above the boiling point of water, nor do we know what the actual upper heat limit for life might be.[5]

Psychrophiles—Cold Lovers

Several of the planets and moons in our Solar System are quite cold. Could alien life thrive in environments of extreme cold? We know from studying creatures on Earth that there are several plants and animals with antifreeze chemicals allowing them to live in the extreme cold. These chemicals suppress intracellular ice crystal formation by super-cooling, which allows them to survive temperatures as low as −40 degrees F (−40 degrees C).

My favorite examples of cold-loving animals are Antarctic fishes that produce chemical compounds with powerful antifreeze properties to depress the freezing point of the fluids just as antifreeze in a car's radiator prevents the fluid from freezing until much lower temperatures are reached. In fact, the survival of these fishes rests on several different antifreeze molecules, called *glycopeptides*, found in all their body fluids, except for their urine. When these glycopeptide molecules absorb the tiny seeds of ice that may form in the blood, they prevent the ice crystals from growing larger. The kidneys of the animal prevent the glycopeptides from entering the urine (in which they would exit the body), thus eliminating the need for resynthesis of the molecules.[6]

In addition to the cold-loving fish, there are various other Earthly creatures adapted to life in the cold. These include *boreal* (cold, Northern Hemisphere) woody species that freeze but use extracellular spaces to accommodate the formation of ice crystals.[7] Many organisms other than plants have also evolved methods of resisting freezing, particularly creatures in Antarctic environments. Hundreds of species of bacteria, protozoa, and algae live on the pack ice surrounding Antarctica. *Obligate psychrophiles* actually require coldness to survive. Their optimum growth temperature is around 50 degrees F (10 degrees C), and they do not survive if exposed to 68 degrees F (20 degrees C).

Cold-adapted life-forms such as photosynthetic eukaryotes like *Chlamydomonas nivalis*, *Chloromonas* (*Scotiella*), *Ankistrodesmus*, *Raphionema*, *Mycanthococcus*, and certain dinoflagellates (small marine creatures) are often visually apparent because they color snow. If someone dared you to eat the snow, you probably wouldn't have much to fear. We appear to be relatively safe from psychrophilic infection since all animal pathogens (disease-causing microbes) appear to be *mesophilic bacteria*, that is, bacteria whose optimum growth occurs between 68 and 113 degrees F (20–45 degrees C).

Some land organisms can withstand periods of unfavorable conditions because they can hibernate for a long time. As we wonder about the possibility of life in outer space, and the possible transmission of life from one planet to another, this ability could be useful for organisms hitching rides on ice comets or other frozen debris. In 1997, various images from a NASA satellite suggested that the Earth is bombarded every day by thousands of house-sized snowballs that break up high above Earth's surface and send down a gentle cosmic rain. If this controversial finding proves true, it suggests that Earth has been continually gaining water and may have acquired several ocean's worth over geological time.

However, even if microbes can survive in the icy heads of comets or in chunks of ice on the ground, frozen microorganisms cannot survive forever. Radiation, either from radioactivity in rock or from cosmic rays falling from the sky, will damage DNA and over millions of years kill the microbes. Even at very cold temperatures, a very low level of metabolic activity would be required to repair DNA and replace old amino acids. Nevertheless, there are many interesting examples of bacteria that have stayed alive for eons in the Siberian permafrost. These formidable survivors have been doing practically nothing for 3 million years, and they've been doing it at 15 degrees F below 0 with no sunlight, no air, and no fresh food. Even at 24 degrees F (–4 degrees C), the bacteria still have a metabolism. Rather than reproduce, they sit on the threshold between life and death, as eons rush by and civilizations rise and die. If such creatures naturally arose on Earth, there is the possibility of life-forms on various worlds with polar ice (for example, Mercury, our Moon, and Mars) and on worlds covered with large amounts of ice (for example, various moons of Jupiter, Saturn, and other outer planets, as well as comets and asteroids).

We can learn a lot about cold-loving organisms by studying creatures living in subsurface Antarctic lakes. The NASA scientist Dale Andersen explored the bottom of Lake Hoare, Antarctica, after drilling holes through the ice. There he found mats of algae similar to the primitive life that dominated the Earth 3 billion years ago. In the surrounding land he also found algae, bacteria, lichens, and mosses clinging to life in a valley sufficiently cold and dry to mimic Mars. If there is life on Mars, or a fossil remnant of life from that planet's earlier, wetter history, it may resemble these Antarctic colonies. As we learn more about Antarctic life, NASA can use this information to better design robot spacecraft that search for Martian life.

There are also several Antarctic organisms in regions with short-term freeze/thaw cycles and diurnal temperature fluctuations of up to 72 degrees F (40 degrees C). One such creature is *Heteromita globosa*, a heterotrophic flagellate.[8] At least 24 species of protists (unicelluar organisms with nuclei) and some fruticose lichens and mosses also grow in these conditions. Many organisms found in the Antarctic sea ice that normally grow at 28 degrees F (–2 degrees C) do not do well when the temperature is raised above 36 degrees F (2 degrees C). Some of the cold lovers die when brought to room temperature because their membranes can't withstand the heat. In the stable low temperature of the sea, the flagellates have lost the ability to synthesize some membrane components, perhaps the lipids or fatty acids, which makes it impossible for them to grow at higher temperatures.[9]

Acidophiles—Acid Lovers

In his novel *Close to Critical*, the American writer Hal Clement discusses aliens that live in sulfuric acid. Could aliens really live in acid so strong it would burn our skin at the slightest touch? Until very recently only four Earthly organisms, all eukaryotes, were known to grow under extremely acidic conditions, near pH 0. These include *Cyanidium caldarium* and three fungi, *Acontium cylatium*, *Cephalosporium* sp., and *Trichosporon cerebriae*. This doesn't mean that the interiors of their cells are acidic. In fact, the Cyanidium maintains its internal milieu at close to neutral pH. In 1995, two acidophilic prokaryotes were discovered: *Picrophilus oshimae* and *P. torridus*, both thermophilic archaeans found in Japan, in *solfataras*, natural steam vents in which sulfur gases are mixed with hot vapor.[10] Living at just slightly more moderate pH values, there are numerous Earthly examples of protistan and fungal organisms, for instance those that thrive in the stomachs of some animals.

The relative acidity or alkalinity of a solution is indicated by the pH scale, which is a measure of the concentration of hydrogen ions in solution. Neutral solutions have a pH of 7.0. A pH of less than 7.0 denotes acidity (an increased hydrogen ion concentration), and above 7.0 alkalinity (a decreased hydrogen ion concentration). Many important molecular processes inside living cells occur within a very narrow range of pH. Therefore, cells must regulate internal pH levels in order to live. The pH may differ locally within an organism; however, most tissues are within one pH unit of neutral.

Although our emphasis is on eukaryotes in this section, I'd like to comment on a few acidophilic bacteria. Most bacteria grow in a range of near-neutral pH values, between 5.0 and 8.0, although a few bacterial species have adapted to life at more acidic or alkaline extremes. For example, when coal seams are exposed to air through mining operations, the pyritic ferrous sulfide deposits are attacked by *Thiobacillus ferrooxidans*, which generates sulfuric acid and lowers the pH to 2.0 or even 0.7. This organism can tolerate high concentrations of iron, copper, cobalt, nickel, and zinc ions and acidity as low as pH 1.3. Many bacteria thrive in acidic bogs, pine forests, and lakes that are between pH 3.7 and 5.5. The acid lover *Sulfolobus acidocaldarius* has a high tolerance for acid strong enough to dissolve your skin in seconds.

Alkalophiles—Alkali Lovers

Lake Nakuru, an African soda lake with a pH of about 10, is home to numerous life-forms. Lake Nakuru supports millions of flamingos that feed on cyanobacteria, such as spirulina, that grow in the lake. Each day the flamingos contribute to the lake 15.6 tons dry weight of fecal and urinary matter, which feeds a standing crop of noncyanobacterial prokaryotes. Twenty different *heterotrophic* (bacteria requiring an organic source of carbon—see note 8) protist species and three rotifer (many-celled aquatic invertebrates) species also thrive in the lake.[11]

Halophiles—Salt Lovers

Could an alien life-form exist in extremely *saline*, or salty, water? The level of salinity is the amount of dissolved salts that are present in water. Naturally occurring waters vary in salinity from almost pure water devoid of salts (for example, snow melt) to saturated solutions in salt lakes such as the Dead Sea with a concentration of salt at approximately 332 parts per thousand. The extreme salinity of the Dead Sea excludes most animal or vegetable life except halophilic (salt-loving) bacteria. Fish carried in by the Jordan or by smaller streams when they are in flood die instantly. Apart from the vegetation along the rivers, the plant life consists mainly of halophytes, plants that grow in salty or alkaline soil. Extreme halophiles, such as members of the genus *Halobacterium*, show optimum growth at 20 to 30 percent salt and disintegrate if this salt level is reduced. Such bacteria are found in the Dead Sea and even on salted fishes and hides.

Supersalty waters support a wide variety of single-celled creatures such as diatoms and flagellates. The African soda lakes such as Lake Nakuru are also supersalty and can support a very diverse microbial population if there is ample food. There are also numerous species of halophilic and halotolerant algae.

The major problem facing halophiles is control of their osmotic pressure—pressure caused by water flow in and out of a life-form. Without such controls, they might lose water to the surrounding environment. The natural tendency is for water to flow from regions where its concentration is higher to regions where its concentration is lower. I have experimented with marine worms whose response mechanism allows them initially to swell when dropped into water with low salt concentrations, and to shrink when dropped into water with high salt concentrations. If the salinity changes are not great, the worms can regulate the water flow in an attempt to return their bodies to normal volume. A much smaller organism, *Dunaniella salina*, synthesizes high concentrations of intracellular glycerol to balance the external osmotic pressure.[12]

Barophiles—Pressure Lovers

Aliens could conceivably live under extreme pressures. On Earth, we know that ciliates can withstand cycling between one to three atmospheric pressures every minute without ill effects, and there are abundant eukaryotic communities on the continental shelf at least to a depth of 6,560 feet (2,000 m). Because there are complex vertebrates in the deepest oceans, it is likely that heterotrophic eukaryotic microbes are present at even great depths, given sufficient food. On the other hand, many surface organisms cannot survive great pressures: Amoebae, for instance, become progressively less able to form pseudopodia (protrusions) with increasing pressure. When amoebae are subjected to a walloping pressure of 6,000 pounds (2,720 kg) per square inch, they actually turn into tiny spheres and are completely motionless.[13]

In 1997, scientists at the Japan Marine Science and Technology Center in Natsushima Yokosuka, Japan (see note 8), found a new species of marine worm living at 21,000 feet (6,500 m) near the bottom of the Japan Trench in the western Pacific, where the pressure is 650 times that at sea level. The worm's body is transparent, offering an easy view of its internal organs. Other wonderful marine worms, living at a mere 1,640 feet (500 m) deep in the Gulf of Mexico, were also discovered in 1997. These ice worms are a part of a previously unknown deep-sea ecosystem that includes mushroom-shaped outcroppings of methane hydrate, a kind of ice that forms in sea-floor mud. The environment is potentially very poisonous because propane, ethane, sulfides, and crude oil seep into the water in this region. To scientists' surprise, hundreds of flat, eyeless, pink worms were found swarming on the methane outcroppings.

Aridophiles—Dryness Lovers

The driest Earthly regions can also support eukaryotic life: Lichens grow on stones and in the Negev Desert, and organisms also inhabit dry, sandy Antarctic valleys where liquid has probably not fallen for 2 million years.[14]

What Does It All Mean?

In this chapter, we've discussed life that survives seemingly at the edges of the possible. On Earth, bacteria and their kin can live in extreme environments and are capable of the most unusual metabolic feats. Some creatures are happy at over 212 degrees F (100 degrees C) in water under pressure; others can survive freezing and remain active if the water remains liquid. Life can thrive under intense pressure on the ocean floor, in saturated salt solutions, in acidic and alkaline environments, and in environments without the slightest trace of oxygen. Some creatures can tolerate these environments; others actually require them. When it comes to diet, there are life-forms that metabolize iron, sulfur, hydrogen, acids, oils, and even stranger things. Life exists even when concentrations of nutrients, obtained from the air or from distilled water, are very dilute.

Bacteria and their kin seem more invulnerable than any aliens dreamed up by science fiction authors. Bacteria can thrive in nuclear reactors. (Even insects are often quite resistant to radiation.) Bacteria can live in vacuums, as evidenced by bacteria recovered from a camera returned by the Apollo 12 crew after three years in a hard vacuum. Bacteria that divide into two without a distinguished parent have a claim to immortality, and bacterial spores can certainly survive for centuries and possibly for millennia.

To conclude this chapter, let me review and summarize current scientific thoughts on extremophiles. The first life-forms on Earth were probably archaea, not bacteria. Charles Darwin suggested that life evolved in a warm soup of organic molecules, but archaea were born in hell: boiling sulfurous pools or hot, mineral-laden, deep-sea volcanic vents at temperatures above the boiling point of water. The common ancestor of all life on Earth may have been a hyperthermophile with a metabolism based on inorganic material, such as carbon dioxide or hydrogen sulfide. Paradoxically, extremophiles have it "safe": The rain of meteorites and asteroids during Earth's early history may have had little effect on the deep-sea volcanic vents or subsurface sea-floor cracks.

In recent years, microbiologists and geologists have probed the very deep limits of life. They have found organisms trapped almost 1.8 miles (3 km) beneath the state of Virginia for millions of years, microbes living off bare rock and water 0.9 miles (1.5 km) beneath the Columbia Plateau, and signs of life subsisting on glass and mineral-rich water beneath the mid-ocean ridges. Isolated in the depths for millions of years, microbes have adapted to their impoverished life with exotic metabolisms and, in some cases, very slow rates of reproduction. These strange and abundant life-forms are shaping our world and its waters to a degree we are only beginning to determine. They may control the chemistry of our planet like an invisible, alien army—silent yet omnipresent. If humankind ever perishes through nuclear war or natural catastrophe, these tiny creatures of the deep will survive. Perhaps Jesus was right when he said the meek shall inherit the Earth.

WEIRDER
WORLdS

Time is a relationship that we have with the rest of the universe; or more accurately, we are one of the clocks, measuring one kind of time. Animals and aliens may measure it differently. We may even be able to change our way of marking time one day, and open up new realms of experience, in which a day today will be a million years.

—George Zebrowski, *Omni*

Heaven and earth are large, yet in the whole of space they are but as a small grain of rice. How unreasonable it would be to suppose that, besides the heaven and earth which we can see, there are no other heavens and no other earths.

—Teng Mu, thirteenth-century Chinese philosopher

If you think the alien bodies, senses, and environments described so far in this book were weird, you've seen nothing yet! In this chapter we discuss life on even stranger worlds, ranging from high-gravity brown dwarfs to the voids of space in the final days of the universe.

From Now to Eternity

What is the ultimate fate of the universe, and what aliens might survive during the final days? Before discussing the aliens, let's first review what the universe will be like in the far future. Fred Adams and Gregory

Laughlin, two astrophysicists, have written a fascinating review article in the April 1997 *Review of Modern Physics* describing the birth and death of the cosmos, beginning 1 million years after the Big Bang and ending an incredible 10^{100} years later. (This scientific notation denotes 1 followed by 100 zeros; our universe today is a mere 10^{10} years old.) In the current scientifically accepted scenario, our current star-filled cosmos will eventually evolve to a vast sea of subatomic particles as stars, galaxies, and even black holes fade.[1]

The death of the universe unfolds in four acts. In our early era, the energy generated by stars drives astrophysical processes. Even though our universe is 10 billion to 20 billion years old, the vast majority of stars have barely begun to twinkle. Stars shine by fusing hydrogen nuclei at their core, forging helium and heavier elements. Massive stars burn brightly but die fast. Stars as heavy as the sun live for about 10 billion years. Low-mass stars have not even begun to evolve.

In about 10 trillion years, the emissions of the lowest-mass stars will revive fading galaxies, temporarily boosting their brightness. Alas, even these last surviving stars will die after 100 trillion years, and star formation will be halted because galaxies will have run out of gas—the raw material for making new stars. At this point, the stelliferous, or star-filled, era draws to a close.

During the second era, the universe continues to expand while energy reserves and galaxies shrink and material clusters at galactic centers. *Brown dwarfs*, objects that don't have quite enough mass to shine as stars do, linger on. Gravity will have already drawn together the burned-out remains of dead stars, and these shrunken objects will have formed super-dense objects such as white dwarfs, neutron stars, and black holes. Eventually even these white dwarfs and neutron stars disintegrate as a result of the decay of protons.[2]

The third era—the era of black holes—is one in which gravity has turned entire galaxies into invisible, supermassive black holes. Through a process of energy radiation described by the astrophysicist Stephen Hawking, black holes eventually dissipate their tremendous mass. This means a black hole with the mass of a large galaxy will evaporate completely in 10^{98} to 10^{100} years.

What is left as the curtain closes on the black hole era? What fills the lonely cosmic void? Could any creatures survive? In the end, our universe will consist of a diffuse sea of electrons. The cosmos may expand forever if the density of matter is too low for gravity to stop the expansion.

I always wonder about the possibility of life in the fourth era, this Dark Era beyond 10^{100} years. Certainly, aliens that depend on water and organic compounds have vanished, but there may be a network of structures spread out over unimaginably large distances, and these organized structures could store information. According to the astrophysicist Gregory Laughlin, these structures, made out of whatever materials are available, will have extraordinarily low energy and will unfold extraordinarily slowly, but in some sense, the structures may always continue to exist in the universe. Could these structures be living? What would the lives of these "Diffuse Ones" be like?

A Day Will Be a Thousand Years

In the Bible (2 Peter 3:8), we come across the statement "One day is with the Lord as a thousand years." During the "Final Days," the consciousnesses of the Diffuse Ones (low-energy-structure creatures) arises from a sea of diffuse electrons or other particles. Their thought and communication processes might appear extremely slow to us in our time frame. But the universe operates on different scales of time: The time frame of the Earth is much slower than that of humans; the time frame of an insect that lives only 24 hours is much faster than that of humans. Similarly, it would not matter to these beings that their thinking, evaluating, and communicating processes were extremely slow by our standards. Their time frame would be proportionate to their speed of thought. The Diffuse Ones wouldn't care that it took them a million years to scratch their "noses" or to wait for their toast to pop up.

The lives of the Diffuse Ones might be primarily ego-intensive. During the Final Days there would be little outside stimuli to which they could react, and they would look inside themselves for entertainment, ideas, and excitement. They may form complex social groups if communication with other, far-away Diffuse Ones is possible.

Even if they progressively slow down, they could run a virtual world in their minds much as a computer runs a program, and the Diffuse Ones would not perceive any slowdown. This means that although the physical universe is a black emptiness of electrons, neutrinos, and leptons, a rich virtual universe could unfold from within.

It is possible that the Diffuse Ones would comprehend that in the past there were accelerated ways of thinking. To overcome their limitations, they could evolve further in the post-black-hole era and choose to organize small pockets of the universe. They would use their thoughts to gather sufficient particles together and cause a small cosmic egg to form. They could ensure their new universe had an influx of free electrons during its existence. In their new universe, conditions might be favorable for rapid thought processes compared to the universe in which the Diffuse Ones were born.

Could a Diffuse One have intelligence and consciousness and not be truly "alive"? Some might argue that Diffuse Ones couldn't be alive if they could not reproduce themselves, given the limited amount of material available during the Final Days. On the other hand, certainly the nonreproducing bacteria living miles under the Earth are considered lifeforms. The definition of life is always difficult. If we posit that life transforms energy and replicates its structure, then even fire is alive.

These discussions remind me of Frederik Pohl's *The World at the End of Time,* in which a number of stars are accelerated at nearly the speed of light for millennia until they are far, far away. To the people living around the stars, only a few thousand years have passed because of time-dilation effects (time runs slower on high-velocity objects relative to low-velocity objects), but in the meantime the rest of the universe deteriorates to a very low energy state with little structure. This is one way of keeping hope (and structure and life) alive in a dying universe. Perhaps some far-sighted beings could toss a few galaxies on a programmed path sending them far away at near light speeds and then bring several back every million years when the rest of the universe has decayed.

Michael Michaund, the former deputy director of the Office of the International Security Policy in the Department of State and a SETI enthusiast, speculates on what humanity (or aliens) could do to avoid eventual destruction:

> Organized intelligences of the universe might avert destructive collapse or dissipation of the universe by isolating controlled regions of the universe from the rest of space-time and universal

evolution, by transferring themselves to another point in time, or by escaping this universe, perhaps to another, younger one.

The physicist Freeman Dyson in his paper published in the 1979 *Reviews of Modern Physics* examines life in the past and notes that it takes about 10^6 years to evolve a new species, 10^7 years to evolve a genus, 10^8 years to evolve a class, 10^9 years to evolve a phylum, and less than 10^{10} years to evolve all the way from the primeval slime to *Homo sapiens*. If life continues in this fashion in the future, it is impossible to set any limit to the variety of physical forms that life may assume. What changes could occur in the next 10^{10} years to rival the changes of the past? Dyson believes that in another 10^{10} years, life could evolve away from flesh and blood and become embodied in an interstellar black cloud, as presented in Fred Hoyle's science fiction story *The Black Cloud*, or in a sentient computer, as presented in Karel Čapek's *R.U.R.* In Hoyle's black cloud, a large assemblage of dust grains carries positive and negative charges. The cloud organizes itself and communicates with itself by means of electromagnetic forces. Dyson notes, "We cannot imagine in detail how such a cloud could maintain the state of dynamic equilibrium that we call life. But we also could not have imagined the architecture of a living cell of protoplasm if we had never seen one."

Some physicists suggest that the amount of energy in the universe will asymptotically approach but never actually reach zero. Life and civilization could get by forever on the tiny residue of energy. However, given an arbitrarily large period of time, the energy available to life-forms may become arbitrarily small. At some point life may encounter quantum mechanical effects. What happens when quantization leaves only one energy state? At this point, it may be possible to exist via the vacuum fluctuations,[3] where the overall universe's energy is arbitrarily close to zero but locally there may be variations allowing transient existence—not so different from what we all have now, transient existences.

If you believe that only flesh and blood can support consciousness, then life would be very difficult in the Final Days when the universe expands and cools, and does not contain water nor much energy. But to my way of thinking, there's no reason to exclude the possibility of nonorganic sentient beings in the final diffuse universe. I call these beings Omega creatures. If our thoughts and consciousness do not depend on the actual substances in our brains but rather on the structures, patterns, and relationships between parts, than Omega beings could think. If you could make a copy of your brain with the same structure but using different materials, the copy would think it was you.

Sometimes I mourn the fact that the ultimate fate of the universe involves great cold—or great heat if there's sufficient gravity to draw all matter together in a single point in a final Big Crunch. It is likely that *Homo sapiens* will become extinct. However, our civilization and our values may not be doomed. Our heirs, whatever or whoever they may be, may find practical ways for manipulating time as they launch themselves throughout the Galaxy. They will seek their salvation in the stars.

And when the stars die, beings like Diffuse Ones could inherit the knowledge and emotions of human civilizations. They may be stepping stones in our final salvation.

Brown-Dwarf Priests

We've been discussing the possibility of life in a future so distant that stars no longer shine. However, even today stars are not necessary to support life or produce light. For example, light may be emitted by chemical processes on a planet far away from a sun. A more intriguing idea is the possibility of life on brown dwarfs—warm planetlike objects far away from suns and therefore without sunlight.

A brown dwarf is an astronomical object that falls midway between a planet and a star. Brown dwarfs have a mass less than 0.08 of the Sun's mass, and their surface temperatures are below 3,900 degrees F (2,200 degrees C). Sometimes described as failed stars, brown dwarfs probably form like stars when interstellar clouds contract into smaller, denser clouds. Unlike stars, however, brown dwarfs do not have sufficient mass to generate the internal heat that in stars ignites hydrogen and creates thermonuclear fusion reactions, the source of stellar powerhouses. Though they generate some heat and some light, brown dwarfs also cool rapidly and shrink. Brown dwarfs look like high-mass planets and may be distinguishable from planets only in their formation mechanism. A brown dwarf is formed directly from a collapsing gas cloud—a stellar process—rather than from the accretion of dust and gas that gives birth to planets.

How could life evolve and survive on bodies with no sunlight? I've already discussed Earthly life that survives quite happily without light, and the fact that the first life-forms on Earth may not have needed light at all. However, although there is no "visible" light available to life on brown dwarfs, warm dwarfs glow brightly in the deep infrared, and this might be exploited by organisms, both for vision and photosynthesis, the manufacture of carbohydrates by plants. While photosynthesis as we know it would be impossible without sunlight, a different form of energy capture could take place in the absence of sunlight. Moreover, lightning discharges that may have played a role in chemical evolution on Earth would be present on brown dwarfs to provide an abundant energy source.

The nearest life beyond the Solar System may not be on a planet orbiting a star but on one of these lonely bodies not married to any sun. Scattered throughout the universe are probably countless bodies of this sort, possibly with water, and ripe for some form of life. It's difficult for astronomers on Earth to detect brown dwarfs because they are intrinsically faint objects. However, in 1995 a cool brown dwarf named Glise 229B was discovered by scientists using the 60-inch (154 cm) Palomar telescope specially fitted with a coronagraph, a device usually used to study stars. Tiede 1, a dim object in the Pleiades star cluster, is also generally accepted to be a young brown dwarf. Numerous other possible brown dwarfs have been discovered since then. A dwarf 10 times the size of Jupiter would produce the right amount of heat to permit liquid water.

What strange biologies might develop in the absence of the light in the violet-to-red range on brown dwarfs or other dark worlds? Creatures could "see" using vibration and electrical sensors such as exhibited by the electric eel and mormyrids, elephant fish that have a brain larger in proportion to their body weight than that of man. Creatures in the dark might also sense pressure differentials. As an example, consider the fishes that live in caves on Earth. They can orient themselves, like most other fishes, by their lateral line sense, moderated by an organ that runs along their sides and registers the surrounding environment by sensitivity to its differential pressures.

The lateral line organs of fishes, which are grooves in their sides, function as distance receptors. Bundles of nerves in these grooves register the differential pressures of the surrounding water and thus enable the fish to perceive objects within a certain range. This kind of sensory organ might also serve a creature that lives on land or in the atmosphere by registering differential air pressures. Perhaps these creatures would also evolve some kind of communication by controlling ambient air pressure, for example by shooting puffs of air at different velocities. Sophisticated communication by tactile means is also possible. Consider the case of the author Helen Keller, who learned to interpret human messages normally conveyed through sight, sound, and speech through a system of direct touch. Braille is a typical example of Earthly communication through finger touch, and we can easily imagine our letters of the alphabet conveyed through 26 touch points on the body.[4]

Aliens on dark worlds might develop a very keen sense of temperature and use this for both communication and exploring their environment. While humans can sense gross changes in temperature, some animals on Earth posses thermal sensors far more sensitive than ours. For example, the mosquito can register differences of as little as one five-hundredths of a degree C at a distance of 0.4 inch (1 cm). Some fishes—the sole is one—respond to temperature changes in the water of as little as 32.05 degrees F (0.03 degree C). The bedbug can crawl along a wall of a bedroom, sense a tiny area of exposed skin, and jump to it.

Humans sense *relative* temperatures. We know that one glass of tea is hotter than another. But we can't say precisely how hot it is. Other creatures on Earth sense absolute temperature. For example, some fish can be trained to recognize a particular temperature within 1.8 degrees F (1 degree C), irrespective of whether the fish came out of a previously warmer or colder environment. Some birds have the ability to maintain their nests at a precise temperature; they make small alterations to the nest if it becomes a degree too hot or cold.

To better understand possible alien worlds, consider the analogy of a painting. When we see a painting we see many different hues not seen by a color-blind animal. (Incidentally, many mammals are color-blind.) Just as we see reds, greens, blues, and all the shades in between, aliens on dark worlds may "see" the world with distinct temperatures. An alien with this ability would perceive and give labels to 100, 101, and 102 degrees in the same way we perceive different colors and name them red, purple, and maroon. Their Leonardo da Vinci might hang a plate with different temperature regions evoking the same emotions as our Leonardo does with his *Mona Lisa*. Their traffic lights might be hot, warm, and cold instead of red, yellow, and green. Their sexy magazines might arouse them with thermal profiles in the same way that a Marilyn Monroe photo in *Playboy* can excite and titillate.

What would happen if we could visit aliens on a dark world and shine an ordinary flashlight at them? Perhaps the aliens would fear a bright light, if they could perceive it, making light a symbol of great evil or holiness. Do brown-dwarf priests dream of white light in the same way theologians conjure indescribable visions of God? Can creatures dream of things beyond their sensory capacity?

Grace Under Pressure

The absence of light in the violet-to-red range would not be a major problem for the development of life on dark worlds. Creatures could develop eyes or photoreceptors that function in the infrared range. These receptors may vaguely resemble our eyes, or they could be structurally very different—for example a collection of hairy filaments, textured skin patches, or heat-sensitive membranes that we would never associate with vision.

Such structures exist on Earth. A snake has a reflector-shaped membrane in a pit beneath each of its eyes. Each membrane consists of 150,000 nerve cells sensitive to heat in an area in which a human would have only three heat sensors. Even if a pit viper is blinded, it has no trouble striking out at a mouse several feet away using its thermal receptors. Like a camera, the snake's heat detectors can reveal the outline of a creature that may only be a fraction of a degree warmer than its environment. An alien who had similar abilities would have a clear view of the recent past via thermal "sight," much as the sense of smell reveals past information. If an alien with advanced thermal receptors stared at my bed 30 seconds after I left it, he or she would still see me.

On high-gravity worlds with little or no sunlight but with soil,[5] there would probably be a greater number of soil-living species than above-ground species. Perhaps there would be many more varieties of worm-like creatures than exist on Earth, and also more burrowing species like our mice, rabbits, and moles. There might be even larger animals underground. Whereas our foxes create dens only to rear young, animals on dark alien worlds might spend the bulk of their time underground.

If high gravity causes soil compaction, then animals need stronger methods for digging. The digging creatures might look like armored moles. On Earth there are examples of social moles that form hives. On a high-gravity world, social alien moles may form hives clustered around sources of heat or subterranean roots.

Certain life-forms could not easily exist on very high-gravity worlds. Ants can lift hundreds of times their own weight, but I doubt ants could flourish on a world with 100 times our gravity. Nevertheless, some insects can survive brutal forces. Insects like the click beetle (*Athous haemorrhoidalis*) average 400 *g* when "jackknifing" into the air to escape predators (1 *g* corresponds to the force we feel on Earth). This means that they can accelerate and decelerate very quickly, causing extreme forces on their bodies. Studies show they can endure a peak brain deceleration of 2,300 *g* by the end of the movement! Aquatic creatures might flourish on high-gravity worlds, although they probably would not develop technology.

Science fiction is rife with creatures from high-gravity worlds. My favorites appear in Hal Clement's *Mission of Gravity* (1954), a novel describing the high-gravity planet Mesklin and its intelligent, insectlike inhabitants. The planet Mesklin is enormous and dense, and it rotates in about 18 minutes, which squashes it flat like a discus. Its atmosphere is hydrogen, its seas methane, and the gravity varies from a comfortable 3 *g* at the equator to nearly 700 at the flattened poles. The Mesklinites (Figure 4.1) are armored centipedelike creatures with a naturally intense fear of heights because falling can be fatal. In the novel, humans land near the equator of Mesklin and are interested in the poles, where only the Mesklinites can go because of the crushing gravity. In one poignant scene from the book, when a human lifts the Mesklinite protagonist from the ground, "All his hearts beat out of sync for fear."

The foot-long Mesklinite evolved from aquatic, jet-propelled ancestors. The Mesklinite has 18 pairs of legs, each of which ends in a suck-

4.1 *A Mesklinite, an insectlike inhabitant of a high-gravity planet, from* Mission of Gravity. *(Drawing by Brian Mansfield.)*

erlike foot, enabling the Meslinite to tightly grip surfaces. Forward pincers function as hands, while a rear set is used for anchoring the creature. Four eyes surround a mandiblelike mouth. Possessing no lungs, a Mesklinite absorbs hydrogen directly from the air through pores. An internal siphon system, originally used by remote ancestors for submerged propulsion, allows them to speak in voices pitched from very low to ultrasonic (high frequencies outside our hearing range).

The Mesklinite culture near the planet's North Pole has all the phobias you would expect on a high-gravity world, and Clement reveals these during an odyssey to rescue a United Planets probe that crashes at Mesklin's South Pole. Barlennan, the Mesklinite, is prepared to endure psychological stress caused by being six inches above ground, or having a heavy object above him briefly, in order to broaden his intellectual and emotional horizons.

Robert L. Forward also describes life on high-gravity worlds in his book *Dragon's Egg* (1980), which describes life on a neutron star named Dragon's Egg, and the compression of time for its unusual inhabitants.[6] The neutron star has the mass of a star but the radius of a small asteroid, so its gravitational field is 70,000 million times that of Earth. In *Dragon's Egg*, the gravity is so high that the atmosphere is only a few micrometers thick. Mountain ranges are about 0.4 inch (1 cm) high.

One can imagine life evolving on a neutron star in the same way as life evolved on Earth. However, the nuclei making up the biological matter do not have electrons bound to them, as they do on Earth. Instead, neutron star biochemistry depends on nuclear reactions mediated by the strong force of the nuclei and not on the electromagnetic forces responsible for terrestrial chemistry.

On Dragon's Egg, the dominant form of life are "Cheela" (Figure 4.2), intelligent creatures with the same biological complexity as human beings and a similar number of nuclei. Their flat, sluglike bodies 2 inches (50 mm) in diameter and 0.2 inch (5 mm) high consist of complex molecules of bare nuclei. They don't have the strength to extend themselves more than a few millimeters above the crust because of the crushing gravity. Similarly, they do not breath or talk because the "atmosphere" is only a few micrometers thick. They communicate by strumming the crust with their lower surfaces. The star's 14,000-degree F (8,000 degrees C) surface radiates sufficient long-wavelength "light" to enable Cheela to see. To Cheela, the surface looks like a bed of glowing coals.

In Chapter 1 we discussed the fact that some of the most successful life-forms on Earth have bilateral symmetry: Only one plane of symmetry divides the animal into approximately symmetrical halves. In contrast, the Cheela do not have bilateral symmetry and can move on their treads equally well in all directions.

The plants on Dragon's Egg make food by extracting energy from the crust through their root system and emitting their waste heat to the cold sky. It is never dark, so the life-forms have never evolved sleep. There is no moon, so the creatures have no months. Dragons Egg does not orbit a star, so they have no year.

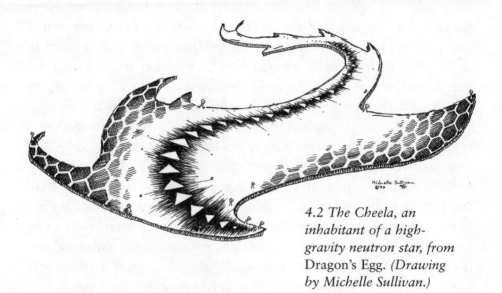

4.2 *The Cheela, an
inhabitant of a high-
gravity neutron star, from*
Dragon's Egg. *(Drawing
by Michelle Sullivan.)*

Obviously, technology on a neutron star would be very different
from Earth's. The high gravity constrains buildings to be quite low and
sturdy. An extremely high magnetic field tends to elongate objects
along the lines of the magnetic field, and moving things across the
magnetic lines is difficult. Cheela find it easy to move east and west,
but difficult to move north and south.

Let's imagine you are able to send a small robotic probe equipped
with an infrared video camera to a Cheela village. As the durable robot
crawls and beams back video to you, you notice that the Cheela do not
have lamps, candles, or electric lights, because there is no dark and no
cold. Even the insides of holes and caves are brightly illuminated by the
glowing warm walls. You order your robot to enter a Cheela home and
discover that the Cheela don't have hanging pictures, hinged doors or
windows, leafed books (because pages would tear if turned), rooftops,
or tops to buildings—all because the gravity is too high. Look up:
There are no airplanes, balloons, or kites. Look around: There are no
whistles, fans, straws, or perfume because there is no atmosphere.
There are no umbrellas, baths, or flush toilets because there is no rain
or equivalent of water. Look at the walls of the Cheela home. Cheela
art is made with fluorescent materials or liquid crystals.

Someday we may find life on neutron stars, although it would be stranger than we can imagine. If star creatures did exist, they would probably not discover us. They would find it too difficult to travel in space. The collapsed matter making up their bodies would transform into normal atoms when the creatures lifted off into a region of low gravity, and the creatures would literally blow up. Because their biology depends on strong nuclear forces instead of electromagnetic forces—and nuclear reactions happen faster than chemical ones—star creatures would live a million times faster than we do. It would be difficult to communicate with such creatures. It would even be difficult to study them with machines, as described in the scenario with the robotic probe. We would have to communicate with them by messages sent by computers. Even if we liked one another, we could never visit them and they could never visit us. The gravity on a neutron star would destroy us, and our gravity would destroy them. We could only enjoy each other's philosophies from afar.

Light Without Suns

In several of the preceding sections, we've been discussing life without sunlight. There are all kinds of potential worlds ready for life without sunlight, including moons circling brown dwarfs. However, the absence of sunlight does not necessarily mean that there is no visible light. For example, on Earth there is an exotic source of light that shines thousands of feet below the ocean's surface. The light may be sufficient to power photosynthesis on the ocean floor—which would provide our first example of photosynthesis divorced from the sun's rays.

This dim light comes from vents that spew out volcanically heated brines laden with metals and caustic compounds. At first, scientists attributed the light to thermal radiation emitted by the 662-degree F (350 degrees C) water, much as hot metal can glow. However, measurements suggest that thermal radiation alone cannot explain the light.

Cindy Lee Van Dover, a marine biologist at the University of Alaska at Fairbanks, first found evidence of the light in the late 1980s, while studying a seemingly blind species of shrimp, *Rimicaris exoculata*. These shrimps swarm around hydrothermal vents on the volcanically active mid-Atlantic ridge that forms part of an underwater mountain range.

Oceanographers had named the shrimp *exoculata* ("lacking eyes") because it appeared to lack eyes, but Van Dover and her colleagues discovered that the animal actually has vision organs, just not in the usual place. Instead of eyes attached to its head, *R. exoculata* has evolved oversized light-sensing patches on the back of its shell.

The shrimp probably uses these eyes to see light coming from the hydrothermal vents—light in the form of a very faint glow recorded by digital cameras and photometers. It's unlikely the light is produced by thermal radiation because it has the wrong frequencies and is more intense 4 inches (10 cm) above the vent, where the water is cooler. Although scientists are not sure of the light's source, there are several possibilities, including crystalloluminescence (produced by the crystallization of dissolved minerals as the hot water turns cold), triboluminescence (caused by the cracking of minerals), or sonoluminescence (caused by the collapse of microscopic bubbles). Further research is needed to be sure that the light is of the proper wavelengths for the organisms to use, either for vision, for photosynthesis, or for phototaxis (movement toward the light). Incidentally, phototaxis may help vent bacteria move toward the chemical nutrients they need for survival.

R. exoculata needs to feed on chemosynthetic bacteria near the vents, but if the shrimp gets too close, it gets cooked. Perhaps the shrimp can use the vent lights to help orient at a safe distance from the hot plumes. The eyes of the shrimps have enlarged retinas that are very tightly packed with photosensitive pigments to capture as many photons as possible in the animals' light-starved environment. Scientists have tried to capture a few of the shrimp for testing, but unfortunately the light shining from submersibles to locate the shrimp blinds them instantly.

Just as exciting is the possibility that deep-sea bacteria use vent light for photosynthesis. This phenomenon already exists: The most sensitive photosynthesizing organism known today is a green sulfur bacterium that lives 262 feet (80 m) below the surface of the Black Sea. Capturing the pale-blue rays of the sun that filter down from the surface, these Black Sea bacteria thrive on a trickle of light measuring only a thousand photons per square centimeter per second.

We've already discussed possible energy sources for life, such as sulfur used by the vent bacteria, and light from stars or other luminous objects. One prerequisite of life is a flow of free energy. Although we think of sunlight and chemicals as energy sources on Earth, alien life could conceivably thrive on other forms of electromagnetic radiation such as infrared light and X-rays, streams of charged particles, heat differentials, and nuclear energy.

The amount of energy concentrated in a particular region of space, and the rate of molecular mixing, places constraints on the evolution of alien life-forms. For example, the rate of chemical reactions between molecules widely scattered in outer space, or in gases, or even solids, may be too slow for sufficient complexity to develop over time. Liquids (or dense gases) may be better suited for chemical reactions. Life may evolve on weird worlds and in strange states of matter, but if I were a gambler betting on the existence of life in the universe, I'd rather place my bet on a world with liquid than on a world without.

5

ORiGIN OF
ALIEN LIFE

Among all the great discoveries of the last five hundred years, to me, at any rate, the biggest, most marvelous discovery of all is the discovery of how life came into being—the discovery which we associate with the name of Darwin and DNA. Two hundred years ago you could ask anybody, "Can we someday understand how life came into being?" and he would have told you, "Preposterous! Impossible!" I feel the same way about the question, "Will we ever understand how the universe came into being?" And I can well believe that the evidence that we need is right in front of us, right now. We just have to look in front of our noses.

—John Archibald Wheeler

The essential building blocks of life—amino acids, nitrogen-bearing heterocyle compounds and polysaccharides—are formed in space. These compounds occur in large quantities throughout the galaxy.

—Fred Hoyle and Chandra Wickramasinghe

Panspermia

To best understand the possibility of life on other worlds, it is important to understand how life may have started on Earth. The origin of life is the most fundamental and the least understood of biological problems. It is central to many scientific and philosophical problems, and to any consideration of extraterrestrial life.

I personally do not believe that the origin of life results from a supernatural event beyond the descriptive powers of physics and biochemistry. Rather, I believe that life arose on the early Earth by a series of progressive chemical reactions starting from molecules present on Earth or from molecules brought to Earth by objects such as meteorites. The idea that Earthly life got help from outer space gained favor toward the end of the nineteenth century, when the Swedish chemist Svante August Arrhenius suggested that Earthly life arose from *panspermia*, a process whereby microorganisms or spores waft through space by radiation pressure. However, today we know that it is unlikely any microorganism could be transported by radiation pressure to the Earth over interstellar distances without being killed by the combined effects of cold, vacuum, and radiation. Arrhenius believed that air currents or volcanic eruptions had carried life spores aloft from their native planet and then electric forces had moved them free of the atmosphere. Since light exerts a very weak pressure, Arrhenius further reasoned that the pressure of sunlight would send these spores far into space.

The panspermia proposal was elaborated in 1954 when J. B. S. Haldane of Britain named the traveling spores "astroplankton" after the Earthly equivalent of plankton—the drifting, microscopic life in the oceans: "One of the earliest parties to land on the moon should be able to look for astroplankton, that is to say, spores and the like, in the dust from an area of the moon which is never exposed to sunlight."[1]

Haldane felt that the astroplankton could survive better in the shade unexposed to prolonged periods of solar radiation. Not only did he think that spores could be carried from one part of the galaxy to another by light pressure, he also felt it was possible that they were "launched into space by intelligent beings."

Various scientists have since argued that even bacterial spores of the type that survive being boiled would be killed as soon as they left our atmosphere. The astronomer Carl Sagan, while at the University of California at Berkeley, calculated that such spores could not even survive the journey from the Earth to Mars, owing to lethal ultraviolet light from the sun and other stars. This hazard would be reduced in the vast space between stars, but cosmic rays (high-speed particles) would pose additional hazards. In spite of this, it *is* possible that microorganisms can survive quite lengthy journeys in space if conveyed within protective rocks. (The precedent for microbial survival in rocks was discussed in Chapter 3, which mentioned Earthly microbes discovered deep beneath the ground in terrestrial rocks at depths of several miles.)

We know that asteroids striking the Earth can displace material into space, and some pieces of Earth may eventually fall on Mars. Similarly, Martian rocks can find their way to Earth. It is conceivable that microorganisms could be transferred between planets this way. Although bacteria in *small* meteorites would die as their rocky vessel burned to a crisp in Earth's atmosphere, a medium-sized meteor would be braked gently by the atmosphere, would not get too hot in its core, and would hit the ground relatively softly. Bacteria riding inside these might survive such a landing. We do know that the Murchison meteorite that fell in Australia in 1969 contained dozens of amino acids (the basic building blocks of proteins), including many that are common in Earthly organisms.

Directed Panspermia

In 1973, the Nobel Prize–winning physicist Francis Crick and Leslie Orgel elaborated upon Haldane's panspermia proposal, suggesting that distant creatures *purposely* sent out spores to different worlds. They called the process *directed panspermia* and went a step further, proposing that the spores had been sent in an "unmanned" spacecraft to avoid damage from lethal ultraviolet radiation or other sources.

What would motivate aliens to seed other worlds with spores? Perhaps aliens desire life to continue when their parent stars die. Perhaps any beings facing imminent destruction prefer to pass on their life via directed panspermia—particularly if they were unable to directly communicate with other worlds by other means.

For the first few billion years on Earth, life was single-celled, as evidenced by fossils. Does this rule out the possibility that complex multicellular organisms such as jellyfish, mice, or ants were sent to our world via directed panspermia? Or is it possible that alien "mice" were sent only to die, staggering helplessly along our "alien" shores, releasing the millions of bacteria in their guts upon their demise. Or perhaps alien cultures sent billions of microorganisms of all varieties, for example, a cocktail of anaerobes, thermophiles, psychrophiles, acidophiles, alkalophiles, halophiles, and barophiles (discussed in Chapter 3), hoping that at least one of them would survive on a far-away world. Given a billion years, these primitive forms may have evolved into multicellular life-forms.

If aliens were aware of planets that contained oceans with organic material, a single bacterium dropped into such an ocean might be sufficient to seed it for life. All life on Earth appears to have a common ancestry, as shown by the universal similarity of its basic chemistry. DNA contains the basic hereditary information of living cells, which all function using the same genetic code. ATP (adenosine triphosphate) transports energy within all cells, from human to bacterial. Additionally striking is the fact that nature has chosen the same 23 amino acids to construct the proteins of life—despite the fact that chemists are aware of numerous other possible amino acids. Furthermore, although the amino acid structures can exist in two mirror-image forms, only the lefthand form is used by living cells. However, all of these clues do not prove panspermia happened, because over the billions of years, natural selection could have eliminated all other less efficient versions of codes and chemicals.

Even more radical than Crick and Orgel's directed panspermia is *continuous panspermia*—Sir Fred Hoyle and Chandra Wickramasinghe's hypothesis that microbes (mainly viruses) are continuously reaching the Earth in debris shed by comets. They suggest that this rain of microbes is responsible for the world's great epidemics. This idea seems far-fetched to me because pathogens usually have great specificity. If fish viruses rarely infect humans, could an alien virus infect us? Viruses evolve to attack particular receptors on cells, and to use the host's cellular machinery. For example, rhinoviruses (the cause of the common cold) seem to have evolved with humans and seem to be specifically designed for infecting humans—so it is unlikely that they just happened to drop in from space able to infect us.

This leads to an interesting question. Is there the slight chance that aliens could infect us? Could aliens, even if coming with only peaceful intentions, do what the Europeans did to the Native Americans when they came to America bringing viruses to which Native Americans had no immunity? One chance of infection arises if all life in the galaxy were related as a result of spores carried in comets and meteorites. Under these circumstances, there's a slight possibility the viruses aliens could bring would be devastating to humanity. If this were true, then our viruses would probably be equally deadly to them. However, if life arose independently on Earth, then it is extremely unlikely that a virus from an alien would have any effect on humans.

Another slight possibility is that alien bacteria could harm us. While viruses are very host-specific, some bacteria and fungi can affect many different animals because they produce toxins (which is what happens with botulism) or kill through mechanical effects such as when a mass blocks a blood vessel. The spiral bacteria of the genus *Leptospira*, for example, can kill cows, dogs, and humans by congesting blood vessels.

Hollywood screenwriters and science fiction authors have sometimes profited from promoting the idea of our microbes infecting alien life. Recall the demise of Martian invaders by Earthly microbes in H. G. Wells's *War of the Worlds:*

> Scattered about, some in their over turned war-machines, some in the now rigid handling-machines, and a dozen of them stark and silent and laid in a row, were the Martians—dead!—slain by the putrefactive and disease bacteria against which their systems were unprepared; slain, after all man's devices had failed, by the humblest things that God, in his wisdom, has put upon this earth.

Science fiction literature is rife with alien organisms that invade human bodies. For example, in Robert Heinlein's novel *The Puppet Masters,* horrible sluglike creatures attach themselves to people, feeding on their bodies and taking control of their minds. Though these stories are fun to read, they have biological weaknesses. Most parasites are rather choosy about the hosts they attack. As mentioned, parasites and their hosts evolve together and are coadapted. Despite the delicious opportunities for Stephen-King-esque stories, this means that alien parasites will not invade our bodies.

Stay away from sushi, not alien flesh.

Sir Fred Hoyle's Black Cloud

In the last section, we discussed Sir Fred Hoyle's theory of "continuous panspermia." Sometime around the end of World War II, Hoyle began to wonder about the large diversity of organic molecules being identified in the dust clouds of the galaxy. Did they suggest life elsewhere in the galaxy? His speculations led to his writing the novel *The Black Cloud* (1957), in which such molecules become organized into a living entity— a black cloud—that heads straight for the Sun, seeking the Sun's energy for nourishment. Unfortunately for Earthlings, the black cloud blocks out the Sun's light, thereby freezing to death a quarter of the world's population. In Hoyle's novel, astronomers are able to communicate with the cloud and warn it that certain aggressive governments have sent hydrogen bombs toward it. In response, the cloud reverses the courses of the bomb-equipped missiles, causing further devastation on Earth. The cloud then departs without exacting further retribution.

Could a life-form like the black cloud really exist? Hoyle's black cloud is vast and intelligent, containing a large amount of interstellar hydrogen. The cloud, 93 billion miles (150 million km) in diameter, has a complex central neurological system consisting of patterns of molecular chains forming brains. The brains are surrounded and interconnected by the cloud's electromagnetic flow and circulating gases that provide energy for the brains and waste removal. When the cloud approaches the vicinity of a sun, it assumes a disklike shape that enables it to absorb energy more efficiently. By condensing hydrogen in a small area of the cloud and producing a fusion reaction, the cloud creates an explosive jet of gases allowing the cloud to move through space.

In order to reproduce, the black cloud finds a nebula of dense hydrogen gases that does not already have an intelligence. The cloud starts to grow a few extra brains inside itself, organizing the magnetic flows and energy storage systems necessary to support intelligence. It then seeds the nebula with the units, extra food energy, and molecular chains. This seed becomes the nucleus of the baby black cloud, which takes the next few million years to mature.

In Hoyle's book there are thousands or millions of intelligent hydrogen clouds rambling about our universe. They can't stay in one area too long, because they can run out of energy and die—especially in interstellar space. They also can't stay a long time near a sun, because a sun's large gravitational forces cause the cloud to start to condense into a solid body.

The black clouds live solitary existences with occasional communication among the clouds using radio transmission on the 1-centimeter

(0.39 in.) bandwidth. These long-distance conversations are usually about mathematics, philosophy, and the nature of the universe.

Although many biologists believe that alien life must be organic and carbon-based, physicists suggest that Hoyle's intelligent black cloud, a completely inorganic entity, could exist. The American physicist Gerald Feinberg has speculated that space itself could hold two life-forms, "plasmodes" and "radiobes," the former evolving within suns, the later in interstellar space. Plasmodes develop patterns of organized motion from random collisions of electrons and ions. They are alive in that they are structured. They metabolize (feed on energy), and they replicate (by converting magnetically random particle clusters into nonrandom clusters).

I believe, as does Freeman Dyson, that life will evolve into whatever material embodiment best suits its purposes. As we discussed in the section on Diffuse Ones (Chapter 4), it is possible that life in the remote future is something like Hoyle's black cloud, a large, organized assemblage of dust grains carrying positive and negative charges. While it's hard for modern-day scientists to imagine this form of life, remember that a century ago we could not contemplate the biochemistry of life on Earth.

Dissolved Life-forms

In my own science fiction novel, *Chaos in Wonderland*, I also describe the evolution of diffuse patterns of simple molecules. As one example, I describe beings with an energy-water intelligence called the Leandra. They have no physical being except for the organized chemicals in pools of water. Their molecules are mind. Their weak electrical fields allow them to communicate. If they are allowed to dry out, they die.

In *Chaos in Wonderland* I go into the details of how such a water intelligence might exist simply as an oscillating chemical reaction. In particular, the reactions form conventional transistors, just like the switches of digital computers. An interesting dialog between the two discoverers of the Leandra helps illustrate the specifics of the reactions:

> "Do you think they're more like a chemical computer than our brain?" Kalinda asked me as she observed the silent Leandra.

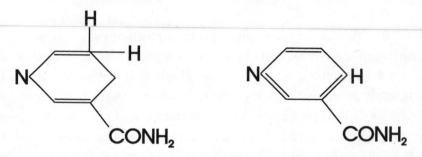

5.1 *Water intelligence based on oscillating chemical pairs of nicotinimide adenine dinucleotide. In* Chaos in Wonderland *these chemical switches act like neurons in our own brain.*

"Our brains *are* chemical computers," I said.

"What about the Leandra?"

"The oscillating chemical pairs from which the water intelligence derives its being are molecules called nicotinimide adenine dinucleotide. This naturally occurring substance exists in a reduced form and an oxidized form in which it was robbed of one of its electrons [Figure 5.1]. In the water, the jump between low concentrations and high concentrations of the two molecules is abrupt and works like a tiny switch. Each chemical switch is like a neuron in our own brains, and the water beings are composed of millions, maybe billions or trillions, of networks of these switches."

This idea is not entirely far-fetched. The use of oscillating reactions to form a chemical computer is discussed in papers recently published in the *Proceedings of the National Academy of Sciences*.[2]

When I fantasize about creatures like the Leandra or diffuse molecular forms such as Hoyle's black clouds, I like to study the chemicals astronomers actually find in dust clouds in outer space. Over the last 40 years, radio astronomers have scanned the dust clouds at infrared and microwave wavelengths and have identified more than 60 organic molecules, including carbon monoxide, water, alcohol, ether, ammonia, acetylene, formaldehyde, and cyanodecapentyne, a 13-atom molecule. The atmosphere of Titan, the largest moon of Saturn, is known to contain at least six hydrocarbons (ethane, propane, acetylene, ethylene, diacetylene, and methylacetylene), three nitrogen compounds (hydrogen cyanide, cyanoacetylene, and cyanogen), and two oxygen compounds (carbon dioxide and carbon monoxide). Titan's atmosphere almost certainly contains other more complex substances. Some scientists believe that the ethane may have fallen to the surface forming an ocean.

In comets researchers have found many possible precursors of life such as methyl cyanide and hydrogen cyanide. In 1986, scientists discovered more than 30 organic (carbon-containing) molecules in Halley's comet, using dust-particle impact spectrometers on spacecraft. The organic molecules were relevant to life and included pyrimidines and purines that are necessary for the message in our genetic codes.

Could life survive on comets? Could living spores survive in the depths of space? In 1983, scientists tried to mimic conditions in space by placing bacteria in a deep vacuum under ultraviolet light. The latter broke up some biological molecules and stimulated the synthesis of others. Most important, some bacteria survived. Given these and other experiments, and the fact that 10,000 tons of cometary dust fall to Earth each year, it is possible that the rain of organic material on our primeval seas played some role in the appearance of sugars and the components of genetic molecules.

Meteorites and Comets

One of evolution's most fascinating enigmas is why Earthly creatures prefer to form their proteins from just one of two types of building blocks. Amino acids, the subunits of proteins, come in two mirror-image shapes that have identical chemical compositions but differ from each other much as your left hand differs from your right. In fact, when amino acids are created in a laboratory, the batch invariably contains equal numbers of left- and right-handed molecules. Presumably the same was true inside Earth's primordial ooze. So why did life favor the left-handed form? Why are proteins in your body made from just one form? These questions remain unanswered. If life originates from nonliving chemicals, there seems to be no convincing reason for one amino acid form to be selected and not the other.

In 1997, researchers discovered that the predominance of one amino acid shape isn't unique to life on Earth: It also shows up in meteorites dropping from outer space. Much of the recent research focuses on a meteorite that fell to Earth in 1969, near Murchison, 80 miles north of

Melbourne, Australia. The Murchison meteorite is a *carbonaceous chondrite*[3]—any stony meteorite containing material associated with life (e.g., hydrocarbons, amino acids, and forms resembling microscopic fossils)—and is generally believed to be a remnant of a spent comet. The meteorite contains 55 amino acids that have no terrestrial counterparts. Eight of the 23 amino acids occur in proteins on Earth. The recent discovery that an excess of one mirror form of amino acid did not evolve on Earth, as many scientists had believed, suggests that the asymmetry may have been the result of chemical processes in the interstellar gases at the time the Solar System was formed. Still, it is possible that the ancient Earth contained equal amounts of the two amino acid forms, labeled L and D, and evolution eventually resulted in the dependence of most organisms on the L form. It is also possible that before life began on Earth, the chemical soup already contained mostly L amino acids, and organisms evolved to use that form.

The unequal amounts of L and D forms in the meteorite demonstrates that natural processes can produce asymmetry in the cosmos. According to the chemists John Cronin and Sandra Pizzarello of Arizona State University, such asymmetry may have been created by polarized starlight that catalyzed the preferential synthesis of L amino acids in the interstellar clouds that became our Solar System. On the other hand, advocates of panspermia have cited the prevalence of the L amino acids as additional evidence for life on comets, because they do not feel there is compelling evidence for nonbiological process to have produced this asymmetry. They argue that the organic matter such as that found on the Murchison meteorite could have played an essential role in nudging life down its left-handed path.

No matter what the source, the new findings may make it difficult to discriminate between organic compounds produced by Earthly organisms and those produced by aliens elsewhere in the Solar System. Researchers have long hoped that the preponderance of L amino acids would be a fingerprint for Earthlings. But if the chemical asymmetry was laid down before the evolution of all life, then aliens may bear the same fingerprint.

Additional analyses of the Murchison meteorite show what appear to be purines and pyrimidines (components of genetic molecules) as well as various hydrocarbons. These and other prebiological substances in comets, meteorites, and celestial dust grains may have been synthesized in solar nebulae before the Earth was formed. Walter Sullivan in his excellent book *We Are Not Alone* notes, "While the seeds of our ancestors may not have fallen from the skies, the most basic components of living things seem to have done so."

Even if complex organic molecules did not fall to Earth from the heavens, we know that sugars and amino acids are easy to create just by shining ultraviolet light in a flask containing a mixture of carbon dioxide, ammonia, and water vapor—a combination resembling Earth's primitive atmosphere billions of years ago. Many scientists suggest that our original atmosphere lacked oxygen as evidenced today by certain primitive forms of bacteria that are killed by oxygen, such as those causing gas gangrene or tetanus.

If our flask of gases is also rich in methane and hydrogen, like the atmosphere of Jupiter and Saturn, and an electrical spark is shot through it for a week, a wonderful array of life chemicals is produced. The water in the flask actually turns deep red and contains alanine and glycine (amino acids), lactic acid, acetic acid, urea, formic acid, glycolic acid, and more. Other gas experiments have produced components of nucleic acids. Merely shining ultraviolet light onto formaldehyde, a molecule assumed to have been generated in the primordial atmosphere, produces ribose and deoxyribose—sugars of RNA and DNA. Researchers have also easily produced ATP (adenosine triphosphate), the energy molecule of all life-forms. (Although the primitive atmosphere may have had much less hydrogen and more carbon dioxide than in these experiments, and was thus a more oxidizing atmosphere, such oxidizing atmospheres also easily yield the chemicals of life.)

I find it quite exciting that out of all combinations of atoms that *might* have been produced by various historical experiments of mixing simple molecules like methane, carbon dioxide, ammonia, and water, the ones most *easily* produced are life's building blocks such as amino acids, sugars, fatty acids, purines, and pyrimidines.[4]

What these experiments suggest is that no extraordinary circumstances were needed to catalyze the diversity of life we see around us. Moreover, if life on Earth could evolve from ordinary processes, then it is likely that humans are not alone in our own Solar System. Other life-forms, perhaps quite primitive, should exist.

The ease with which the basic building blocks of life are produced in the simplest of experiments once prompted the Nobel Prize–winner Melvin Calvin to write, "We can assert with some degree of scientific confidence that cellular life as we know it on the surface of the earth does exist in some millions of other sites in the universe." The Nobel Prize–winner Christian de Duve wrote, "Life belongs to the very fabric of the universe. Were it not for an obligatory manifestation of the combinatorial properties of matter, it could not possibly have arisen naturally."

The First Life-form?

On Earth, nucleic acids, such as DNA (deoxyribonucleic acid) and RNA (ribonucleic acid), contain the basic genetic information of all life-forms. This information is expressed as a sequence of four different chemical bases. RNA molecules are thought by some to be the most primitive "life" forms that first evolved: They spontaneously fold into complex structures, and they reproduce, given the right conditions. Today we know that the folding patterns of RNA affect their function and survival in the presence of enzymatically or biochemically harsh conditions.

If complex spores did not arrive on Earth on comets, the period of chemical evolution on Earth—during which organic compounds gradually accumulated in the primitive seas—probably began about 4,000 million years ago. Hydrogen cyanide (HCN) is central to most of the reaction pathways leading to abiotic formation of these simple nitrogen-containing organic compounds. (*Abiotic* refers to processes that do not involve living components.) HCN is readily formed by reactions such as

$$2CH_4 + N_2 \rightarrow 2\ HCN + 3\ H_2$$

and

$$CO + NH_3 \rightarrow HCN + H_2O.$$

HCN is the precursor of organic molecules such as purines and pyrimidines, which make up molecules such as DNA and RNA.

Many researchers suggest that RNA was the original protogene, the first informational macromolecule, and the first structure at the threshold of life. Today, researchers are attempting to induce RNA strands, in a suitable environment, to reproduce themselves and undergo adaptation by evolution. The genetic information of many viruses is encoded in a single-stranded RNA molecule.

Scientists have long wondered whether proteins came before nucleic acids or vice versa. Proteins seem to be made only on instructions from nucleic acids, yet nucleic acids cannot perform without the help of catalytic proteins. Perhaps the manner in which proteins and nucleic acids interact evolved from a simpler and different process. For example, we know that nucleic acids can multiply without help from proteins. RNAs can act as enzymes, snipping RNA molecules into parts that can then recombine. Perhaps during early evolution on Earth, RNA could not only reproduce itself but could evolve through replication errors, setting the stage for the evolution of more efficient DNA and RNA systems. In the lab, it is possible to create RNA helices and double-stranded RNA simply by mixing nucleotides and phosphates (RNA building blocks) in a flask exposed to a slowly rotating light source simulating cyles of daylight and darkness. RNA subunits trapped in clay can also link into long chains that reproduce themselves.

Other researchers feel that proteins were originally able to replicate themselves, and then "invented" nucleic acids. When mixtures of amino acids are heated to very high temperatures and the resultant proteinoid material is dissolved in hot water and cooled, they form microscopic spheres that look like certain bacteria. The spheres show many lifelike properties such as the catalysis of chemical reactions, surfaces that resemble membranes, and an ability to proliferate. Some researchers feel that the agent causing "mad-cow disease" may consist only of a protein that appears to proliferate in the brain, and this supports the idea that during the course of evolution, a protein life-form could have preceded life-forms based on nucleic acids.

Silicon Life

So far, we've been focusing on carbon-based life. However, researchers speculate that alien life might be based on chains of silicon atoms instead of carbon chains, as is the case on Earth. According to chemical laws, there are only two elements cable of forming the long kinds of chains we think are needed for life: carbon and silicon. Would it be possible for life-forms even on Earth to be based on silicon?[5] This appears to be unlikely, although a complex system of something akin to organic-chemistry reactions could take place with silicon chains in liquid ammonia instead of water. However, ammonia is liquid only within a narrow range of intensely cold temperatures, making it a less likely environment for life than water. Frozen water is quite remarkable because it is less dense than liquid water, and this causes ice to float to the top of oceans during frigid weather. In a liquid-ammonia ocean, on the other hand, frozen chunks of ammonia would sink, thereby exposing the surface of the liquid ammonia to the cold so that eventually all the ammonia in the ammonia sea would be frozen.

Despite these caveats, not only could life theoretically occur in liquid ammonia with temperatures near −58 degrees F (−50 degrees C)—with weaker bonds involving nitrogen predominating in the metabolism—but life could also occur in hydrocarbons,[6] with a mixture of hydrocarbons functioning as the solvent, a dissolving or dispersing agent. (At some point in their life cycles, most Earthly organisms seem to need at least a minute amount of solvent in order to thrive.) For example, I like to imagine little creatures living in petroleum. Reductive reactions are ones in which an electron donor like hydrogen transfers an electron to another agent. Such reactions, for example, hydrogenation, could be used as an energy source. This is not as far-fetched as it sounds. Many extremophiles on Earth flourish in organic solvents toxic to most other life-forms. For example, certain microbes thrive in toluene, benzene, cyclohexane, and kerosene, sometimes at solvent concentrations of up to 50 percent (the other 50 percent being water). The microbes can be found in soil and deep-sea muds and they degrade crude oil and polyaromatic[7] hydrocarbons. These kinds of microbes may be useful as biodegradable agents that reduce toxic wastes.

Carbon does have some unique characteristics that make it an ideal candidate for leading to life. It can bond to itself in long chains and can form bonds to four other atoms at one time. This allows theoretically for a huge number of different compounds. Note, however, that life could be based on less versatile atoms. For example, it is not necessary for an atom to bond to itself to form long chains. In fact, the chains could be made of two or more atoms in alternation. The physicists Gerald Feinberg and Robert Shapiro have speculated that life could be constructed using an alternative chemistry in which the possibilities are not as vast as those of carbon. For example, although the English language can be communicated and stored using 26 letters, it can also be coded as successfully, although less compactly, with 1's and 0's, the binary code used by computers. In the same way, a less complex chemistry could serve as the genetic basis for life, with a large number of components needed in each molecule or cell.

What Is Life?

We've been discussing the chemical evolution of life and various chemicals that could conceivable make up life. Recently, the MIT chemist Jules Rebek created an organic molecule that reproduces itself—a molecule that Rebek considers a primitive form of life. Whether or not it is truly alive, it is definitely not life as we know it. For example, Rebek's J-shaped molecule is held together by some of the same chemical bonds as proteins, DNA, and RNA—but the molecule reproduces in a chloroform solution. For those of you who are chemists, this primitive life-form consists of an "amino adenosine triacid ester."[8] In a chloroform solution, Rebek's molecules can copy themselves at rates up to a dizzying million times per second.

As a result of Rebek's research, we should expand our concepts of what raw materials are needed for extraterrestrial organic primal soups. Rebek's and other experiments, while not telling us what really did happened billions of years ago on Earth, can give us a clue as to what *could* happen on this world or others in the universe.

Of course, all this talk about other forms of life does not address the question, "What is life?" In fact, the very contemplation of alien life-forms begins with this question. One might identify as life anything that ingests, metabolizes, and excretes, but this description might apply to a car, rust, or a candle flame. Others define life as a departure from thermodynamical equilibrium—but much of nature (like lightning and the ozone layer) is out of equilibrium, and thus, while fitting that definition, is not life. Biochemistry-based definitions of life that require proteins or nucleic acids seem restrictive. For example, if we found an alien worm that could do everything a worm could do on Earth but was made of different molecules, certainly we would not declare it "lifeless." In the final analysis, most definitions may be impractical on alien worlds.

We do have some idea of how quickly life evolved on Earth. The Earth formed through an assembly of ancient "planetesimals," bodies with a radius of around 3 miles (5 km). These chunks began crashing into one another, producing fragments that in some final "Great Bombardment" assembled into the planets of today. On Earth, primitive life originated very soon after the Great Bombardment, which ended about 3.8 billion years ago. Numerous pieces of fossil evidence suggests that primitive life was already well established on Earth 3.5 billion years ago. A study of Earth's geological history suggests that it was much easier for primitive cells to evolve from organic chemicals than for multicelled creatures to evolve from single-celled creatures, because multicelled creatures do not appear in the fossil record until less than 1 billion years ago.

If simple life-forms exist on a planet, what are the chances they will evolve into higher organisms like humans? During the evolution of life on Earth various catastrophes have taken place, such as the one that caused the destruction of the dinosaurs or the one that killed 80 percent of the marine animals during the Middle Cambrian period (about 515 million years ago). Each of these events cleared the Earth for a burst of evolution in new directions. It's unlikely that these chance events were replicated anywhere, so life cannot evolve exactly as it has here on other worlds. However, once the spark of life is ignited, it will, over and over again, flame into whatever crack or niche is available to it, leading to a conflagration of different creatures.

Speculations on the origin of life have been with us for centuries. In times past, scientists believed in spontaneous generation, also called *abiogenesis,* by means of which it was believed that even large creatures develop from nonliving matter. For example, pieces of bread wrapped in rags and left in a dark corner were thought to produce mice, because after a week, there were mice in the rags. Spontaneous generation appeared to explain the appearance of maggots on decaying meat. However, by the eighteenth century it had become obvious

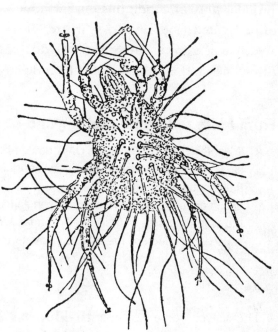

5.2 Arcarus electricus. *In the 1800s, Andrew Crosse believed he had synthesized this microscopic creature by running an electric current through a porous stone soaked with hydrochloric acid and potassium silicate. The life-form, named Arcarus electricus, was shown in H. M. Noad's* Lectures on Electricity *(London, 1849).*

that higher organisms could not be produced by nonliving material. The origin of microorganisms such as bacteria was not fully established until the nineteenth century, when Louis Pasteur proved that microorganisms reproduce.

One of my favorite proponents of abiogensis was Andrew Crosse, who in the early nineteenth century allegedly created organisms in the laboratory using electricity. Crosse reported that when he soaked porous stone in a mixture of hydrochloric acid and potassium silicate, then ran an electric current through the stone, horrifying monsters of microscopic size emerged (Figure 5.2). Today, we assume that the creatures were there to begin with, although unobserved!

Crosse's conclusions may seem far-fetched, but only because he believed *complex,* multicellular organisms arose in a flash from simple chemical manipulations. Many scientists believe that the first shreds of life on the primordial Earth *did* arise from nonliving matter through *biopoiesis,* which means the creation of life from nonliving material that contains the necessary chemicals. According to this theory, during this process molecules slowly grouped, then regrouped, forming ever more efficient means for energy transformation and reproduction.

Under *present* conditions on Earth, new forms of life are not likely to be created from nonliving matter. If life is continuously being created, the new forms are not so well adapted to the environment as existing ones and are thus unable to compete successfully.

From Mars to Europa and Beyond

If we, the discoverers of the DNA double helix, deserve any credit at all, it is for persistence and the willingness to discard ideas when they became untenable. One reviewer thought that we couldn't have been very clever because we went on so many false trails, but that is the way discoveries are usually made. Most attempts fail not because of the lack of brains but because the investigator gets stuck in a cul-de-sac or gives up too soon.

– Francis Crick

The search for extraterrestrial life has driven famous astronomers throughout the ages, from Sir William Herschel (1738-1822), the discoverer of Uranus, to Percival Lowell (1855–1916), instrumental in the discovery of Pluto, to Carl Sagan (1934–1997), who provided valuable insights into the origin of Earthly life. In fact most astronomers are fascinated by the notion of alien life, and many modern astronomers believe that two ice-moons of Jupiter, Enceladus and Europa, are likely candidates for life. This fascinating idea of life on Jovian moons is not new and has been discussed by several authors in the past. For example, Arthur C. Clarke, Richard C. Hoagland, and Dr. Roger Jastrow have long suggested that living forms could exist on Europa, beneath ice-covered oceans kept liquid by Jovian gravitational forces.[9]

I have also speculated about life on Jovian moons, stretching readers' imaginations to the limit with fanciful descriptions of life on Ganymede, one of the most remarkable moons in our Solar System and the setting for my science fiction adventures in *Chaos in Wonderland*. Before scientifically speculating on life on Jovian moons and other planets, I'd like to digress and spend a few paragraphs telling you about the offbeat world, biology, and society of my own hypothetical Ganymedean creatures. In *Chaos in Wonderland* I develop an entire civilization. The creatures and ecology I describe are purely fictional, but one wonders whether such a race *could* evolve on Ganymede, given the right conditions. . . .

Ganymede and the Imagination

I sometimes dream that a hundred years from now a spaceship from Earth discovers the remnants of advanced life-forms on Ganymede, one of the water-containing moons of Jupiter. What happened to them? No one knows. They may have metamorphosed into structures not recognizable as life by human eyes, or may have become spores waiting to be awakened in some hidden crater.

Musings such as these led me to imagine the enigmatic Ganymedean creatures in *Chaos in Wonderland*. The shy, sentient race of creatures known as the Latööcarfians spend their days contemplating intricate mathematical patterns. Status in their society is determined by the beauty of their dream structures.

In my novel, the Latööcarfian civilization develops inside a huge air pocket within the ice of Ganymede. The ceiling of the subterranean air chamber is lined with phosphorescent minerals and bioluminescent (glowing) bacteria that supplement the dim sunlight that penetrates the ice.

The Latööcarfians' bodies are composed of aluminum gallium arsenide with traces of silicon from Ganymede's icy soil. These materials cause their heads to be conductors of electrical signals, and their thoughts resemble the flow of electrons in computer chips. The Latööcarfians therefore think at speeds not achievable by Earthly, carbon-based life-forms (Figure 5.3).

Since gallium arsenide emits light when subjected to an electric current, the Latööcarfians display intricate patterns that sparkle with glittering lights. Like a

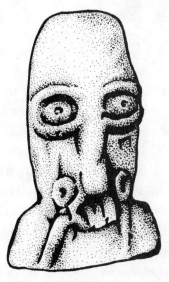

5.3 A Latööcarfian from Chaos in Wonderland. *These brainy mathematicians have semiconductor heads. Status in their society is based on the beauty of their fractal dream structures.*

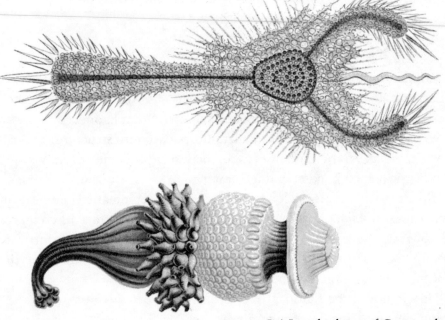

5.4 Ice plankton of Ganymede, from Chaos in Wonderland.

million fireflies dancing to some unheard rhythm, the beautiful head-displays light up the dark Ganymedean evenings. Their blood consists of *electrorheological fluids,*[10] which transform from a liquid to a solid and back again in response to variations in an electric field. *Piezoelectric* materials such as quartz and zinc oxide line their mouths, torsos, and alimentary canals. The oxides expand or contract when they are exposed to a voltage, as their molecules twist to align their internal charges with the electrical field. As a result, these substances act as mechanical devices that curl or extend in response to electrical signals from the Latööcarfians' heads.

In *Chaos in Wonderland,* tiny protozoans live within the ice of Ganymede and form an integral part of the Latööcarfian ecosystem. These planktonic creatures, called öös (Figure 5.4), migrate between the surface of Ganymede and the ceiling of the subterranean Latöö-carfian air chamber in a journey that lasts 17 years. The complex ecology of the öös and dozens of other creatures are discussed in detail in the novel, which also describes the adventures of two anthropologists in the Ganymedean air pocket.

Aliens in Europa

With this fanciful description of strange biologies I developed for the creatures in *Chaos in Wonderland*, I'd like to return to serious contemplation of microscopic life on other moons and planets in our Solar System. Life may seem delicate, yet it arose on Earth under conditions that seem harsh to you and me.

Europa, the fourth largest satellite of Jupiter, has long been considered one of the few places in the Solar System (along with Mars and Saturn's moon Titan) that could possess an environment supportive of primitive life-forms. Five billion years ago, Jupiter was more like a miniature sun than a planet, shedding enough heat to allow Europa's surface to be covered with an ocean rather than ice. Today, Europa is known to contain significant amounts of frozen water. Saturn may also have emitted heat because it is similar to Jupiter in size and composition. There also were other processes that could have given Enceladus, a moon of Jupiter, and Europa open oceans for the Sun to shine on.

Europa is about the size of Earth's Moon and is covered with smooth white and brownish-tinted ice, instead of large craters like so many other bodies in the Solar System. The cracked appearance is probably due to stressing caused by the contorting tidal effects of Jupiter's strong gravity. The warmth generated by tidal heating may be sufficient to soften or even liquefy some portion of Europa's icy covering.

In 1996, images of Europa from NASA's *Galileo* spacecraft gave further support to the idea that "warm ice" or even liquid water still exists today beneath Europa's cracked icy crust. There are places on Europa resembling ice floes, along with suggestions of geyserlike eruptions. The images also suggest that there is geological activity on Europa. In some areas the ice is broken into large pieces that have shifted away from one another, but obviously fit together like pieces of a jigsaw puzzle. This is evidence that the ice crust is lubricated from below by warm ice or liquid water.

In April 1997, newer close-up images of Europa showed a hypnotic array of fine-scale ridges, cracks, and faults on the icy surfaces. Taken by the *Galileo* spacecraft, the pictures confirmed previous findings that blocks of ice on Europa have moved and rotated, as if they had slipped across an underlying layer of warm ice or water. Sunlight shining through cracks in the ice to the top of the water layer may have produced prebiotic, complex organic molecules.

In a typical ice moon we find rock, metal, water ice, dry ice, frozen ammonia, and frozen methane. Eons ago, these materials in vapor or liquid form were exposed to the young Sun's more energetic ultraviolet rays, likely producing various organic compounds on the surfaces of moons such as Enceladus and Europa. If life evolved on these two worlds in the past, there is a possibility it is still alive today in liquid water under their ice crusts. The heat generated by tidal (gravitational) interactions among Io, Europa, and Jupiter appears to be sufficient to melt the ice under the crust of Europa. The *Voyager 2* satellite even observed Enceladus to send out plumes of water. Geothermal energy may sustain life on these worlds in the same way that geothermal heat sources can sustain Earthly life. In my opinion, our first alien visitors will be Europans and Enceladians, transported back to Earth for observation when the United States decides to spend the necessary money to send a robot to these worlds, remove samples from under the ice, and return them to Earth.

Life on Mars

Mars, the fourth planet from the sun, has always been a source of mystery and speculation. Among the most imaginative science fiction involving Mars is the Barsoom series by Edgar Rice Burroughs (1875–1950), the author of the Tarzan books. I recall fondly reading about John Carter's exotic encounters with Martian races and their distinctive languages, customs, environments, and social and political organizations. The Barsoom series incorporated many contemporary ideas of the day, for example, the notion of a dying Mars with dead sea bottoms and canals leading water from the poles as popularized by astronomer Percival Lowell.

Although today we know there are no Martian canals, no one doubts that Mars was once a far warmer and wetter planet. In 1997, studies of Martian meteorites indicated that water flowed at or near the surface of Mars as recently as 700 million years ago. One meteorite contained a large amount of iddingsite, a mixture of clays and iron oxides that forms only in the presence of water. If surface water existed on Mars less than 1 billion years ago, then life may have persisted on Mars for far longer than researchers suspected.

In recent years, there has been increasing speculation about life on Mars. In 1996 the world was stunned by a potato-shaped Martian meteorite, designated ALH 84001, containing possible evidence of primitive life in the red planet's past. Electron microscopes revealed tiny structures resembling terrestrial microbes. Although they are still controversial, some scientists believe they are the fossilized remains of ancient single-celled Martian organisms.

Several pieces of evidence suggest that the structures found in the meteorite result from Martian life. Electron microscope images reveal clusters of elongated shapes no more than 4 millionths of an inch (100 nanometers) long. Looking like minuscule sausage links, these shapes could simply be flecks of mineral. However, they bear a striking resemblance to the earliest, larger microfossils on Earth, which formed 3.45 billion years ago. Dark rims of magnetite (Fe_3O_4) and iron sulfide (FeS) were also discovered. These iron compounds are synthesized by certain bacteria on Earth, particularly anaerobic (oxygen-hating) strains.

Scientists also discovered organic molecules called polycyclic aromatic hydrocarbons, or PAHs, on the meteorite. Ordinarily their presence would not suggest biologic activity—PAHs are observed often in such diverse bodies as meteorites and interstellar clouds, presumably as a consequence of star formation. However the meteororite's PAHs have a distribution resembling what is expected when simple organic matter decays.

The NASA team leader, David McKay, admits that none of these findings by themselves offer definitive proof of primitive life on the meteorite known as ALH 84001. The findings can be mimicked by purely inorganic mechanisms, and there is some controversy over the interpretation of the observations.

The tiny size of the hypothetical Martian microbe fossils also provides an endless forum for debate, as evidenced by dozens of letters written to scientific journals. Is there a lower size limit for microorganisms? Some scientists have argued that "creatures" the size of those found in the Martian meteorite would be far too small to hold the chemical and genetic machinery considered essential for life. At 0.8 to 4 millionths of an inch (20 to 100 nanometers) in length, the Martian forms are at best one hundredth the size of the smallest microfossils of ancient Earthly bacteria ever found. However, the bacteria of the genus *Coxiella* (small, gram-negative pathogens) are as small as 8 × 16 millionths of an inch (200 × 400 nanometers, or 0.2 × 0.4 micrometers), which would be closer to a fivefold difference in size. Thus it is conceivable that even smaller bacteria could have evolved on Mars. The McKay team has also since suggested that many features they see in ALH84001 are bacterial appendages rather than organisms.

What is the tiniest alien we can ever expect to find? On Earth the diameters of the smallest known microorganisms are quite close to the theoretical minimum diameter (0.14 micrometers) for cells calculated from the size of macromolecular components necessary and sufficient for life. Considering a cell packed with its DNA, ribosomes, enzymes, lipids, and so forth, a theoretical minimum diameter of 8 millionths of an inch (0.2 micrometers) is more likely. Note that an oval object with a diameter less than that would contain only about 100 million atoms, an assembly hard-pressed to carry out information storage, metabolic and assembly pathways, and replication processes needed for life.[11]

6

ALIEN SeX

It is only by a quirk of evolution that our sexual organs and excretory organs are united. This has led to the human race seeing sex as dirty and embarrassing. If alien sexual appendages developed in different locations, aliens would not have hang-ups, and their entire sexual psychology would be different.

—Clifford A. Pickover

The earthworm burrowing through the soil encounters another earthworm and says, "Oh, you're beautiful! Will you marry me?" and is answered: "Don't be silly! I'm your other end."

—Robert Heinlein

Bisexuality immediately doubles your chances for a date on Saturday night.

—Woody Allen

Do Aliens Have Sex?

The 1995 science fiction movie *Species* featured sexy "Sil," the unfortunate product of a union between alien and human DNA. If you disregard the implausible genetics, you might enjoy watching Sil as she prowls Southern California looking for men with whom she can mate, before killing them. Sometimes Sil resembles a blonde bombshell; at other times she transforms into a fanged, dripping creature in the tradition of the movie *Alien*.

When pondering the intricacies of creatures such as Sil and other exotic alien biologies, my thoughts inevitably lead to alien sexual practices. Alien sex is not only stranger than we imagine; it is stranger than we *can* imagine. Nevertheless, as with our previous discussions, we can still get an idea of the possibilities by studying Earthly creatures.

In early science fiction literature, evil aliens were often portrayed as spiderlike or reptilian beings who desired to devour or mate with human females. To many of us, trans-species sex seems abhorrent, and we could no more think of being turned on by a reptilian alien then we could by fornicating with an ape. However, alien behaviors may span the gamut of strange desires and relationships. Obviously, sex between aliens and humans could not produce offspring because our genetic makeup would be very different.

On Earth, sex has advantages over simpler forms of reproduction, such as when a single amoeba divides into two parts to produce two creatures. With sexual reproduction, reproductive cells from parents come together and fuse, resulting in offspring that are slightly different from either parent. On the other hand, when offspring develop from cuttings, buds, or body fragments, genetically they are exactly like their parents, as much alike as identical twins. Any major change in environmental circumstances might exterminate an Earthly or alien race practicing this simple form of procreation, since all offspring would be equally affected. When egg and sperm unite, they establish genetic diversity among the population so that some offspring will have unique traits for surviving environmental changes.

In many Earthly animals, sexual differences are apparent both internally and externally. Internally, there is *primary sex differentiation*, into males with testes and females with ovaries. *Secondary sex differentiation* can take many forms, for example, the human male's deep voice and beard and the female's enlarged breasts. The lovely tail feathers of the peacock, the large claw of the fiddler crab, the antlers of a moose, and the powerful body of a harem master in a fur seal colony are all distinctively male characteristics associated with the sexual drive of males. In nature, females are often of comparatively quiet disposition and drab appearance. Their function is to produce and nurture eggs safely and inconspicuously. The male function is to find and fertilize the female, for which both drive and display are helpful.

We are accustomed to thinking of sexually reproducing species as having two sexes—male and female. However, there is no reason to assume that aliens would obey this rule. Even on Earth there are many exceptions to it. For example, the nineteenth-century naturalist Fritz Muller described *Tanais*, a remarkable genus of crustaceans in which the male is represented by two distinct forms. In one form the male carries numerous smelling threads. In the other form, the male has more powerful pincers to hold the female during copulation. Thus, one type of male finds many females but cannot secure them as easily; the other type finds fewer females but hangs on with greater tenacity. The result: Both male forms produce approximately the same number of offspring.

The condition of separate sexes, while common, is not universal. In fact, two sexes within the same individual are typical of the more sluggish animals. Earthworms, land snails, slugs, flatworms, tapeworms, and barnacles are all double-sexed individuals, or *hermaphrodites*. All have ovaries and testes producing mature eggs and sperm at the same time. Nevertheless, cross-fertilization accomplished *between* individuals is common. Self-fertilization, although possible, is usually avoided. Each member of a mating pair of individuals introduces sperm into the body of the other member.

In mammals, it is usually possible to determine the sex of an animal by inspecting primary or secondary sexual features. Although there are sometimes confusing instances of intersexuality or hermaphroditism, these are not common. With invertebrates, the task becomes more difficult. As just mentioned, the *Tanais* crustaceans have three "sexes." The world record holder for most number of sexes in a species is the single-celled *Paramecium amelia,* which is considered to have 8 different sexes. The simple creatures in the genus *Chlamydomonas* have no less than 10 sexes—though to regard these as 5 male and 5 female types (as some have done) seems fanciful.

How many sexes are aliens likely to have? This is difficult to predict. Aliens might have one sex (or no sexes) if they replicate by splitting themselves—like protozoa, which reproduce by simple binary fission without the preceding exchange of genetic material. In "The Trouble with Tribbles," David Gerrold's episode of the *Star Trek* TV show, we find galaxywide furry pets engineered for producing babies and vaguely reminiscent of guinea pigs but with no distinguishing features or even faces. They eat almost anything organic; indeed, they "devote half of their metabolism to reproduction," are "born pregnant," and "reproduce at will." Anyone who has seen the episode remembers the rain of Tribbles down upon Captain Kirk when he opens a ceiling vent. In the course of the story, the tribbles multiply, especially in dark cupboards and closets. They make a purring, trilling sound when stroked, which seems to soothe everybody's nerves. I've not been able to ascertain whether tribbles require two sexes to mate, and would appreciate hearing from knowledgeable readers on this subject.

What are the chances that we could *determine* the sexes of aliens who had more than one gender? Because the external genitals of intelligent Earthly animals are sufficiently distinct to allow an animal to be sexed at a glance, aliens might also be visually distinct. After all, the male and female sexual organs serve different purposes, and form follows function. Also, some of the appearances of sexual organs and secondary sexual characteristics, such as pubic hair and breasts on Earth, probably evolved solely to attract mates by visually differentiating the sexes.

Sometimes even locating a mate can be difficult. For example, both sexes of the West Indian marine fireworm live in crevices on the sea floor but come out to breed. They can find one another only by means of the luminescence they produce, an eerie light visible only in complete darkness. To help locate one another, they emerge about a half hour after sunset when all daylight is gone but before the moon rises, a situation that confines them to a monthly breeding period of three or four days after the full moon. Aliens with a less developed visual sense will use other senses to help attract one another. Perhaps they could even make use of a "staging area" where an assembly of creatures congregate for mating. Unfortunately a local crowd may be an open invitation to predators.

If aliens are anything like some of the invertebrates on Earth, visual identification of the sexes could be very difficult. Some insects require dissection before we can determine their sex. There are also mammals in which the male and female genitals are surprisingly similar in appearance. The spotted hyena *Hyaenidae crocuta* has an almost complete lack of external sexual dimorphism—the external genitals of the

two sexes closely resemble each other: The female's clitoris is perfo-
rated by the urogenital canal and resembles the penis of the male.
Oddly enough, the female also has scrotal pouches beneath the clitoris.

Aliens will likely exhibit all sorts of exotic behavior that we would
identify as sexual deviations or "unnatural" sex acts. If these aliens
visit Earth, their sexual practices will not please conservative religious
groups that already detest what they view as human sexual deviations
or "crimes against nature." However, when considered in perspective,
the "crime against nature" concept makes little sense, because the majority of "deviant" or "perverted" sexual behavior is *commonplace*
throughout the animal kingdom. Animals practice oral sex, rape,
cross-species copulating (the equivalent of bestiality in humans),
sadism, and exhibitionism. Homosexuality is quite common among
animals such as monkeys, bulls, cows, rats, porcupines, guinea pigs,
rams, antelopes, donkeys, horses, elephants, hyenas, bats, mice,
martens, hamsters, raccoons, and dogs. Some animals even change sex
during their lives. If you frown upon homosexuality for religious reasons, how would you feel if we discovered intelligent and kind aliens
who changed sex during their lives. Is it possible that you would feel
uncomfortable with homosexuality in humans but not in aliens?

Some homosexual activities in animals appear to boost reproduction. For example, female cows often mount each other, thereby signaling to any bulls in sight that they are ready to reproduce.
Heterosexual rams with strong sex drives will mount either other
males or ewes. Sixteen percent of domesticated rams never mate with
females during breeding season. Researchers find that dosing pregnant
animals with certain hormones greatly increases the mother's odds of
producing homosexual offspring.

Some animals on Earth *pseudocopulate*—mate with objects, not the
opposite sex of their species. Male bees, wasps, and flies often "mate"
with flowers whose parts resemble those of female insects of the same
species as the males. Masses of pollen become attached to the male insect during pseudocopulation and are transferred to the next flower
visited, thus pollinating it. Perhaps on alien worlds, alien creatures are
compelled or attracted to have sex with plants or other animals.

Alien sex may be vicious if it is anything like the sex of insects. On
Earth, a number of female insects eat their mates during intercourse.

The male praying mantis can copulate successfully when his head has been eaten away by his adoring mate. The females among certain flies gnaw away at their lovers. The most delicious example of carnivorous sex occurs in the fly *Serromyia femorata,* which mate belly to belly with mouth parts in passionate embrace. At the end of mating, the female sucks the body contents of the male out through his mouth. Want more examples? When the male bee introduces his penis into the vagina of the queen, the penis immediately breaks off and he bleeds to death. The sex organs of the bristle worm *Playnereis megalops* are eaten by the pursuing female to facilitate fertilization. The females of certain firefly genera flash in a pattern that attracts males of another genus. When a male comes calling, expecting sweet love, the female quickly devours him. The list is endless. . . .

Sex Change

One of the wildest forms of Earthly sexuality occurs in the worm *Diplozoon paradoxum,* or "paradoxical double animal." This creature happily resides on the gills of carplike fish and, as a hermaphrodite, can mate with itself. Two separate parasitic worms can grow together in the middle of their bodies and become effective Siamese twins joined together until death. The vagina of each half of the hermaphrodite becomes permanently linked to the sperm duct of the other half. This odd arrangement led one prominent researcher, William Bolsche, to term the beast "a love monstrosity, an erotic Briareus with four sexual organs mating crosswise in a double marriage."

Aliens need not stay one sex through their lives. On Earth some birds start with two embryological ovaries, but only the left ovary develops to maturity while the other one remains rudimentary. If, however, the left ovary is destroyed by disease or deliberate experiment, then the right will develop, not into a mature ovary, but into a functional testis. Certain female worms also change into males once they have laid their eggs, and then they will seek out a group of females to mate with. Females can artificially be converted to males by cutting off half of the posterior ring of their bodies. The oyster (*Ostrea edulis*) can change sex with a frequency dependent upon temperature. Warmer temperatures cause more frequent changes.

For hermaphroditic sea bass, mating begins when one fish takes the female role and releases just a few of its eggs; the male-acting fish then ejects sperm into the water to fertilize the eggs. The fish then switch roles, with the former male releasing a few eggs of its own, and the former female getting its chance to play the male.

Perhaps the most famous animal sex change occurs in the wrasse, or cleaner fish, of the family Labridae, which eats parasites from the bodies of other fishes. As reported by R. Robertson in a 1973 issue of *New Scientist*, "The ultimate ambition of female cleaner fish is to become male." Usually, a male dominates a group of females, but if he abandons his harem, the most dominant of the remaining females becomes a male in just a few hours. To be more accurate, the females are really hermaphrodites with small amounts of active (but walled-off) testicular material scattered throughout their ovaries. It's obvious how such quick sex changes can have survival advantages to a species, and it's likely that alien sexuality is just as strange to us as cleaner fish sexuality.

On Earth, sex changes are easy to produce in animals simply by injecting hormones during early stages of development. In the chick, the sex can be changed several hours after hatching. If a female chick is injected on hatching with the male sex hormone, testosterone, it will develop into a fully functional cock. Even when injected at later stages of growth, the male hormone causes growth of the comb, crowing, and aggressive behavior. On the other hand, female sex hormones such as estrogen, when injected into the female, stimulate early growth of the oviduct and feminize the plumage, and when injected in the male suppress comb growth.

Alien Eggs

Dual-gender aliens would not have to require sperm to reproduce. On Earth, the unfertilized egg often possesses all the potentiality for full development. The process of fertilization by a sperm introduces the nucleus of the male sex cell into the female egg, and stimulates the egg to begin development, but these two functions are separate. *Parthenogenetic* development, meaning without benefit of sperm, occurs naturally in various kinds of animals such as the waterflea (genus

Daphnia). Unfertilized eggs of starfishes, sea urchins, various worms, and other marine invertebrate animals can be caused to develop by treatment with a weak organic acid. Unfertilized frog eggs can be stimulated to start developing by gently pricking the egg surface with the tip of a fine glass needle that has been dipped in lymph. Frog eggs developing parthenogenetically become males, since only one X chromosome is present in each cell. A queen honeybee begins her reproductive life with a store of sperm received from a male during her nuptial flight. Throughout spring and summer, almost all her eggs become fertilized and develop into females. Toward the end of summer, when the sperm supply runs low, eggs cease to be fertilized and develop into drones, ready to mate with a new queen should the occasion arise.

In my novel *Chaos in Wonderland*, I discuss a race of translucent, attractive, humanoid aliens, of a genus I called *Rheobatrachus,* who have strange sexual practices. After the male has fertilized the female's eggs with his sperm, the female swallows the eggs and broods them in her stomach. The nurturing female stops eating during the breeding period so that the stomach acid does not destroy the eggs. Actually, the egg capsules secrete a prostaglandin that stops the mom's stomach from secreting hydrochloric acid. The stomach is transformed from a digestive organ into a protective gestational sac! When it is time for the babies to be born, the mother's esophagus dilates and the young creatures are shot from her mouth. Although readers of *Chaos in Wonderland* found this idea hard to swallow, I had to remind them that the idea of a stomach serving as both a digestive and reproductive organ was inspired by the reproductive strategies of certain frogs on Earth that also use their stomachs as reproductive organs.[1]

Alien Penises

Most males of higher vertebrate mammal species have a copulatory organ or penis that also provides the channel through which urine leaves the body. Penislike structures also occur in many reptiles and in a few birds, including ducks. The corresponding structure in lower invertebrates is often called the *cirrus*.

Alien penises or cirruses could be quite strange to look at, judging from the amazing array of penises on Earth. Reptile and insect penises

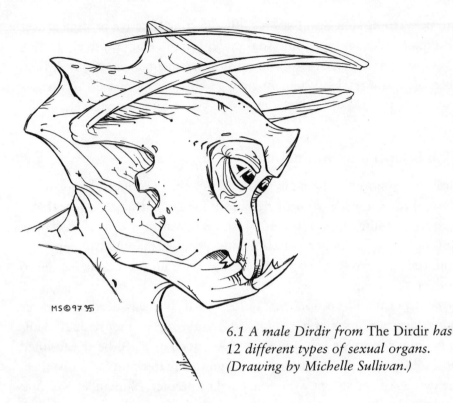

MS©97

6.1 *A male Dirdir from* The Dirdir *has 12 different types of sexual organs. (Drawing by Michelle Sullivan.)*

often come with spines, hooks, knobs, and corrugations to secure the female once the penis is inserted into her vagina. If you try to separate a copulating pair of snakes, the results can be devastating to them. The penis of the male may be ripped off, and the female may be severely lacerated. The spider monkey (genus *Ateles*) has a barbed penis with numerous small black projections.

In the science fiction literature there are several examples of creatures with strange sexual penises. In Jack Vance's *The Dirdir*, the humanoid creatures are tree-dwelling carnivores with cool, flexible skins resembling polished bone. There are 12 different types of sexual organs in Dirdir males (Figure 6.1), and 14 types of female organs. Each type is compatible with one or more types of the opposite gender, and each type has certain traditional cultural attributes. Although a Dirdir's basic gender is obvious from the being's skin color and body size, the exact type of sexual organ is a Dirdir's most closely guarded secret.

On Earth, among the animals with the most penises are various reptiles with paired penises, or *hemipenises*. It is also possible for each hemipenis to be split into two, resulting in four effective protuberances for animals like snakes. Some snakes deploy these alternatively in successive matings, perhaps to allow more frequent copulation. A final piece of snake trivia: When Borneo's yellow-lipped seakrait snakes mate, as many as half a dozen males may pile on a lone female.

Alien Sperm and Intercourse

On Earth, sperm are the male reproductive cells produced by most animals. With the exception of nematode worms, decapods (crayfish), diplopods (millipedes), mites, and just a few others, sperm always have whiplike tails. In higher vertebrates, especially mammals, sperm are produced in the testes. The sperm unites with an egg of the female to produce offspring.

Alien sperm will no doubt be quite varied. If aliens need to transfer microscopic amounts of hereditary material from one individual to another via a spermlike cell, I would expect such a cell to be streamlined and mobile to facilitate transport of genetic material over a distance. The details of the sperm would be hard to predict. Human sperm look like tadpoles, and most mammalian sperm have the same sort of configuration. Even sponges have sperm with whiplike tails. Some sperm, however, are quite elaborate (Figure 6.2). The sperm of a liverwort plant has several tails, and a lobster sperm has three tails. The sperm of a cycad (palmlike plant) resembles a hairy onion. The sperm of the maja (an Asian finchlike bird) and the crayfish look like starfish. Alien sperm may be loaded with strange accoutrements designed to facilitate fertilization.

In a number of Earthly creatures, sperm attain remarkable lengths. In amphibia of the family Discoglossidae, sperm can be as much as 0.08 inch (2 mm) in length, and swimming insects of the genus *Notonecta* have sperm with staggering lengths of 0.5 inch (12 mm)! Certain tiny ostracods (crustaceans) have sperm that are six times as long as their entire bodies. These sperm are curled up compactly as they mature.

Judging from the rich panoply of Earthly organisms, I wouldn't expect the size of the sperm to be related to the size of the alien. On Earth, the size of germ cells in a species is not related to the size of the individuals within that species. This means the largest animals do not

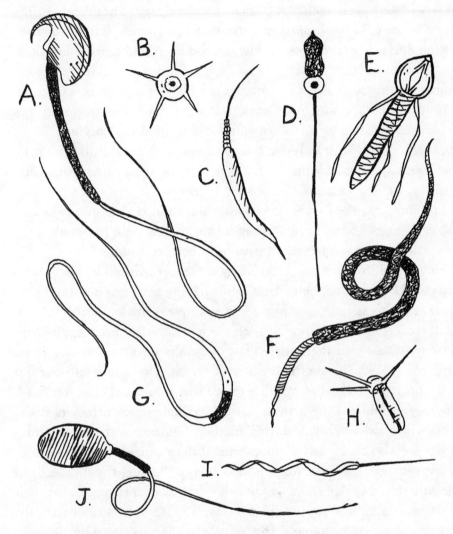

6.2 *Sperm of various animals: A—field mouse; B—maja; C—tortoise;
D—sturgeon; E—anomura; F—raja; G—echidna; H—lobster; I—crow;
J—human. (Drawing by Michelle Sullivan.)*

have the largest sperm and eggs. In fact, the longest mammalian sperm
are found in the Chinese hamster, whose sperm are 250 microns in
length (a micron is one thousandth of a millimeter).

The duration of the alien coital act is hard to predict. On Earth, intercourse can range from a few seconds to many hours. Copulating
snakes can remain in union for almost a day, probably because the

male's spiked penis makes it difficult to disengage. The female of the fluke species *Schistosoma heamatobium* (a parasitic flatworm) lives within a fissure in the male's body and has been said to exist in a state of "permanent copulation." Insects can go for days copulating. For example, members of *Anacridium*, a genus of winged insects, can go an awe-inspiring 60 hours. As in snakes, insect copulations are sometimes prolonged because of the tight interlocking of sexual organs. When copulation is difficult to break off, the penis itself breaks off. The male dies, but happily his dismembered penis remains inside the female acting as a useful plug for the deposited sperm.

In mammals, the longest copulation duration occurs in minks and sables. Matings of the sable lasting for as long as eight hours from the moment of insertion to withdrawal have been recorded.

How would we feel psychologically if we were visited by aliens who offered body parts to their mates while having sex or as a nuptial gift? On Earth this is quite common. For example, a male Australian redback spider offers its entire body to the female. After inserting its copulatory palpus into the female, it somersaults into the female's jaws. The female feeds during the transfer of sperm. The male's self-sacrifice helps prolong intercourse and therefore the number of eggs fertilized. Another example is the male sagebrush cricket, which offers its fleshy hind wings to the female during mating. Sometimes the male's forewings are also consumed. Perhaps one of the strangest body-consumption practices occurs during the matriphagy ("eating the mother") of the Australian spider *Diaea ergandros*—babies of this species eat their mothers as a natural part of their life cycle. After the babies hatch, the mother becomes a "living refrigerator" for her young, who begin to suck on the blood from her unresisting leg joints. After a few weeks, they have consumed her entirely.

Sex in Science Fiction

Let's conclude with some more unusual examples of alien sexuality from the science fiction literature. In my novel *Chaos and Wonderland* I describe some of the most unimaginable alien sexual practices. For example, the Navanax are a race of humanoids about our size with two arms and legs, but that's where the similarity ends. Around each one's body is a flap of sensitive skin mottled with brown, blue, and faint yellow dots. When a Navanax shuffles about the ground, its cloak of colorful flesh looks like an undulating, psychedelic jelly roll. Most of the time the Navanax remains concealed beneath its flesh flap.

There are usually two outcomes when one Navanax meets another. If two Navanax meet face to face, they try to consume each other. If one approaches the other from behind, they have sex. They are hermaphroditic and can be male or female as they choose. In fact, they change sex many times during a sexual encounter. Every Navanax has a penis on the right side of its head. The penis is a few inches behind a genital slit that leads to an ovary. Oddly enough, these descriptions of Navanax mating behavior are stimulated by actual observations of Earthly sea slugs that are hermaphroditic.[2]

For another example of alien sex, consider the Polarians in Piers Anthony's *Cluster*. They are teardrop-shaped creatures about 6 feet (1.8 m) tall when fully extended. A muscular socket at the bottom of their bodies holds a large wheel. The upper part of the body contains a small ball. When two Polarians have the hots for each other and want to make a baby, the male and female circle each other, the male following the seductive scent trail laid down by the female's wheel. They circle faster and faster, spiraling in toward each other until they meet. Their two balls touch to create an electrifying spinning kiss. As they slow down, the male releases his wheel and together the male and female spin it, and in moments it unfolds into a young Polarian.

In the 1950s, Philip José Farmer's stories "The Lovers," "Mother," and "Open to Me, Sister" opened new territory with controversial themes of alien-human sex and incest. In Farmer's novel *Strange Relations*, Mother is a large, intelligent alien with protective outer camouflage resembling a large boulder. She seduces her mates by exuding an attracting musk. The "mate" goes inside Mother and is induced to attack a conception spot on the inner wall. Any abrasion of the conception spot leads to pregnancy and the mate is then devoured by the mouth cavity.

William Tenn satirized biology, sex, and aliens in many short stories, including "Wednesday's Child" (1956), in which a woman gives birth to herself, and "Venus and Seven Sexes" (1949), a treatise on Venusian procreation. David Lake, in *The Right Hand of Dextra*, describes octofugoids (creatures with octopuslike and funguslike characteristics) in which only females are intelligent. The male is very small and photosynthetic. The Dextran reproduces by scattering spores that have been fertilized by male pollen. The spores grow only in warm, moist surroundings, requiring many years of stable climate to reach maturity and intelligence.

As a final example, consider the Azadians in Iain Bank's *The Player of Games*. The Azadians are a hedonistic triple-sexed race: male, female, and "apex." The apex is socially and politically dominant. Males produce sperm. The apex supplies the egg, which it inserts into the female, whereupon she produces a zygote. The apex also has a reversible penis/vagina that can accept sperm.

From Sex to Humor

> Perhaps I know best why it is man alone who laughs; he alone suffers so deeply that he had to invent laughter.
>
> **—Friedrich Wilhelm Nietzsche (1844–1900)**

> Space is almost infinite. As a matter of fact we think it is infinite.
>
> **—Former vice president Dan Quayle**

A tentacled alien from Alpha Centauri writes the number 12345, while his mate slices off his left tentacle. They both break out in laughter. Although we might not find this particularly funny, intelligent aliens may indeed have a sense of humor. What strange reactions analogous to laughter might they have? Do all highly intelligent life-forms eventually acquire something like humor? If they found something funny that we find painful, could we get along?

The researcher Martie Saxenmeyer notes, "Laughter is the meeting of two disparate ideas across a chasm. The discontinuity is what makes the humor. Since we use laughter to defy pain, then I think we could easily understand that an alien might do the same."[3]

We could probably get along with aliens that have different ideas of what's funny—after all, we do this with fellow humans—so long as alien humor doesn't lead to acts of murder and torture. We would probably need to make lists of taboo subjects to avoid—for example, humans can't stand Zetamorph jokes because they involve noises a Zetamorph makes when you kill them . . . Humans don't make puns around Scolexes because it outrages their sense of linguistic symmetry, which they've fought holy wars over . . . Don't make any jokes involving Swiss cheese in front of an Alpha Centaurian because the cheese reminds them of their messiah . . . Don't make any jokes involving farts in front of Vespids because these jokes are somehow disparaging to their mothers . . .

The Swedish xenopsychologist Johan Forsberg believes that "humor is a reaction to the unexpected. Something similar may indeed exist in other curious alien species. Aliens, however, may not exhibit laughter because laughter incapacitates an individual and may be detrimental to survival."[4]

If Forsberg's theory is valid, then some alien jokes would probably be understandable by us. In any case, many species may have jokes, including language puns, body function jokes (fart jokes), and jokes with punch lines. Exchanging libraries of humor with aliens might be the best way to study and understand the differences in our cultures. However, the danger in such a humor exchange is apparent if aliens watch our "funniest" shows whose main themes are people poking one another's eyes, getting hit by iron bars, being burned with hot irons and electrical devices, and tripping down stairs. If humans find humor in all of this and more (for example, see movies such as *Home Alone*, *Pulp Fiction*, and *Fargo*, where audiences laugh at incredible horror and dismemberment), it is not a stretch of the imagination that an alien culture could find the torture of humans very amusing.

A sense of humor is a social instinct that helps us deal with painful and embarrassing situations. Humor is a defense mechanism to protect our emotional systems. What kinds of social and mental defense mechanisms could aliens evolve to make life more bearable? One possibility is that they could develop partial amnesias or false memories, but a sense of humor has greater survival value. Humor may be a natural consequence of empathy, a survival trait for a species, so most species capable of sufficient cooperation to explore outer space would evolve a sense of humor.

In Robert Heinlein's book *Stranger in a Strange Land*, the Martian protagonist does not understand humor, but he finally learns about it by watching monkeys in the zoo, and by studying people's reactions to similar circumstances. The biggest monkey beats on the smaller monkey. The smaller monkey goes over to an even smaller monkey and beats on it. The smallest monkey cries.

Laughter is life's cry of defiance to an uncaring universe.

7 COMMUNICATION

To a frog with its simple eye, the world is a dim array of greys and blacks. Are we like frogs in our limited sensorium, apprehending just part of the universe we inhabit? Are we as a species now awakening to the reality of multidimensional worlds in which matter undergoes subtle reorganizations in some sort of hyperspace?

—Michael Murphy, *The Future of the Body*

I can conceive of no nightmare so terrifying as establishing communication with a so-called superior (or, if you wish, advanced) technology in outer space.

—Nobel Laureate George Wald, Harvard University

The Music of Flowers

What if spaceships from another world suddenly appeared in our skies? What if tomorrow morning you turned on your radio and heard a strange, pulsating tone, and what if you learned that the same thing was happening across our planet?

You are David, the nerdy computer genius played by Jeff Goldblum in the science fiction blockbuster *Independence Day*. Only a day ago, a giant alien mothership arrived in Earth's orbit and immediately began to transmit a cyclic tone down to the nations of Earth. The world has frantically tried to understand the aliens' intentions—until you decipher the alien message: The aliens give Earthlings one choice: Become their slaves, or die. It's a countdown to weapons firing. The President of the United States attempts, unsuccessfully, to reason with the creatures, who demonstrate their massive orbital firepower by destroying large U.S. cities. The military forces of many nations try to retaliate, with little effect.

If we really ever do receive a message from the stars intended to be deciphered by us, just how will it be sent, and how difficult will it be to interpret? If we decided to reply, how would we send a message? One possibility is that we or aliens would use radio waves beamed into space at frequencies between 1 and 10,000 megahertz, because these frequencies travel relatively easily through space and through the atmosphere of planets like our own. The first part of the message would be easy to understand to attract attention, such as a series of pulses representing the numbers one, two, three. This could be followed by more intricate communications.

Could beings on other worlds already be sending beams of electromagnetic waves modulated in some mathematical way, for example in sequences corresponding to the prime numbers? What message would you send to the stars? In the late 1970s, the astronomer Carl Sagan wanted to send *music* on the unmanned *Voyager* satellites destined to leave our Solar System after exploring several planets. Sagan felt strongly that our music was a measure of our achievements and representative of our emotional and intellectual development. Other scientists wanted to send images and information on the chemistry of life on Earth. Eventually it was decided that images, natural sounds, and information could be sent on a phonograph record. But they wondered what sounds and images should be sent to represent a diverse sampling of life on Earth. What would *you* send? Here are just a few of the sounds that were finally sent into space with the *Voyagers*: the songs of the humpback whale, a kiss, a heartbeat, the boom of a Saturn V rocket taking off, frogs, crickets, volcanoes, laughter, and all the languages that humans speak. Images included flowers, trees, animals, oceans, deserts, supermarkets, highways, houses, and humans engaging in all sorts of activities.

Scientists had wanted to send the image of a naked man and a nude pregnant woman holding hands, but this was rejected for puritanical reasons. NASA countered with the suggestion that in its place they could send a photograph of famous statues, such as Michelangelo's David. Scientists rejected this because it might be misinterpreted to mean that any Earthlings who removed their clothes turned to stone. As a compromise, black and white *silhouettes* of the hand-holding couple were included. To help establish the date of the historic recording for an extraterrestrial audience, scientists marked the cover with an ultrathin film of pure uranium-238. By its natural decay over the course of time, this tiny bit of radioactive material would provide a clock for determining the age of the record.

Would aliens be able to decipher the images and sounds sent on the *Voyager* satellite? What if they misinterpreted our intent, and thought the sounds were to be interpreted as images, or the images as sounds! They'd be dancing and snapping their "claws" to "music" from an image of a flower. They'd "look" at the acoustical signals from the audio sections and see something quite abstract, perhaps like paintings by Willem de Kooning or Jackson Pollock.

Deciphering a Message from Tiber

Many science fiction novels have dealt explicitly with alien signals and their decipherment. For example, in Buzz Aldrin and John Barnes's best-selling novel *Encounter with Tiber,* astronomers on Earth detect a signal from Alpha Centauri, the triple star of which the faintest component is the closest star to Earth, about 4.3 light-years away. Scientists first attempt to determine around which of the stars the alien transmitter is orbiting by analyzing the Doppler shift in the waves coming from the transmitter (the Doppler shift is the change in frequency occurring in waves coming from a moving object).

Bits and pieces of the signal seem to be strangely ordered, like a sequence of tones: two different pitches stuttering at an enormous rate. Unfortunately, the Earth's atmosphere is nearly opaque to radio at the transmission wavelength of 315 feet (96 m), because the signal cannot easily penetrate the ionosphere (a region of the Earth's atmosphere in which the number of ions, or electrically charged particles, is large enough to affect the propogation of radio waves). Thus it is impossible to catch more than brief snatches of the message, even using the most sensitive radio telescopes on the ground. Luckily, the scientists find a way to make use of a space station upon which they mount a simple antenna to listen to the signal.

Scientists are skeptical initially about the so-called "signal" and suspect that strange events in the stellar atmosphere of one of Alpha Centauri's stars is somehow causing one star to act as a giant laser at the signal's wavelength. Perhaps a huge electrical storm in the atmosphere of a planet circling those stars could produce such a signal, with the gravity of Alpha Centauri A acting as a lens to focus and amplify radiation. The idea that the signal might be sent by aliens is discounted by many, not just because of sheer improbability, but also because it doesn't seem likely that an intelligent species would try to contact others using radio on a wavelength that would be mostly blocked by the water vapor–rich atmosphere of a planet with life.[1]

Although Alpha Centauri is our nearest neighbor, a spaceship would take about 110,000 years to travel from Earth to Alpha Centauri—time enough for four Ice Ages—assuming we traveled at the speed at which the Apollo astronauts went to the Moon. A radio signal, which is more than 26,500 times faster, takes about four years to traverse the distance.

Despite their skepticism, the scientists continue to study the signal, and discover it is a pattern of high tones, low tones, and silences. Assuming that the silences are spaces, and because the transmission comes as triple beeps, it seems likely that the message is in base 8, meaning that the number system is based on 8 digits (our base-10 system has 10 digits, 0 to 9).

In the story, scientists call the high tones beeps and the low tones honks. There are eight possible combinations of three beeps and/or honks:

beep	beep	beep
beep	beep	honk
beep	honk	beep
beep	honk	honk
honk	beep	beep
honk	beep	honk
honk	honk	beep
honk	honk	honk

The digits were likely to stand for the digits 0 to 7, which are the 8 digits for a base-8 system. The string of digits in the message could represent pictures or text.

The most common numbering system on Earth is base 10. In other words, we have 10 digits, 0 through 9. In our base-10 representation, each number can be expressed as a power of 10. For example, the number 2,010 is $2 \times 10^3 + 0 \times 10^2 + 1 \times 10^1 + 0 \times 10^0$, where $10^3 = 1,000$, $10^2 = 100$, $10^1 = 10$, and $10^0 = 1$. However, there's no reason to assume that aliens would use a base-10 number system, and it's unlikely that a message from the stars would arrive in base-10 numbers. On Earth, our mathematical calculations are based on 10 because of the number of our fingers, our first counting instruments. In fact, our language suggests the connection between fingers and our number system—we use the world *digit* to designate both a number and a finger. Given that our base-10 system comes from our use of 10 fingers, what would a base-8 system tell us about the anatomy of aliens? Perhaps a base-8 system would denote an alien with a thumb and 3 fingers on each hand, or a creature with 8 tentacles, or 1 thumb and 1 finger on each of their 4 arms. An even wilder possibility is that the aliens have 3 heads and these are all the combinations of nodding and shaking that are possible!

As scientists study the message, they find it repeats every 11 hours and 20 minutes. Each group of 16,769,021 base-8 numbers takes about two and a half seconds to be received, so there are 16,384 such groups in all. What could it mean?

The first thing to check is the "Tiberian" number 16,769,021. Does it have any unusual properties? It turns out that you can use a simple factoring program to determine that it is equal to 4,093 4,097—two prime numbers.[2] Since a prime number isn't evenly divisible by another number, an alien could transmit a gridlike pattern whose size is the product of two prime numbers; as a result, there are only a couple of possible arrangements for the numbers in the grid. For example, the pattern could be a photo consisting of an array of pixels like the one on your computer screen. On the other hand, if the image were composed of, say, 10,000,000 pixels with many factors, there would be a very large number of possibles arrangements, such as $5 \times 200,000$, $10,000 \times 1,000$, and many others, and this would make it difficult to decode the image.

In *Encounter with Tiber*, it turns out that the eight groups of honks and beeps represent eight different intensity values in an image: 0 for black, 7 for white, and 2 to 6 for intermediate intensities. By representing these brightness values on a grid of 4,093 × 4,097, the astrophysicists determine that each transmission is a frame of a movie. When played sequentially on a computer, eight creatures are seen waving as they climb into a spacecraft! Other more technical information follows, including instructions on how to find an alien encyclopedia containing poems, paintings, music, literature, science, engineering, and jokes of a civilization centuries in advance of Earth's.

Would you like to view such an alien encyclopedia? In *Encounter with Tiber*, some people on Earth worry that humanity is not ready for advanced knowledge from the encyclopedia. "What if you'd given Napoleon the atomic bomb?" scientists and politicians ask. "What if the Civil War had been fought with airplanes dropping poison gas on cities?" Should the encyclopedia be made available to all the nations on Earth?

Do you think that communication with aliens would create widespread hysteria? The psychoanalyst Carl Jung believed that contact with superior beings would be devastating and demoralizing to us because we'd find ourselves no more a match for them intellectually than our pets are for us. Such fears and jealousies might cause various extremists groups, such as the Ku Klux Klan, to try to kill the aliens.

Mathematical Messages

> We feel certain that the extraterrestrial message is a mathematical code of some kind. Probably a number code. Mathematics is the one language we might conceivably have in common with other forms of intelligent life in the universe. As I understand it, there is no reality more independent of our perception and more true to itself than mathematical reality.
>
> **—Don DeLillo, *Ratner's Star***

Probably the easiest method of establishing communication with an alien species is through mathematics, the universe's Rosetta Stone. Any space-faring or technological race would know about mathematics. It is clear from studying our own history that mathematics has fascinated humans since the dawn of civilization. Has humanity's long-term preoccupation with mathematics arisen because the universe is constructed from a mathematical fabric? In 1623, Galileo Galilei echoed this belief by stating his credo: "Nature's great book is written in mathematical symbols." Plato's doctrine was that God is a geometer, and Sir James Jeans believed God experimented with arithmetic.

Are space-faring aliens also mathematicians? Can we share our math with them? Certainly, the world, the universe, and nature can be reliably understood using mathematics. Nature *is* mathematics. The shape assumed by a delicate spider web suspended from fixed points, or the cross section of sails bellying in the wind, is a catenary, a simple curve defined by a simple formula. Seashells, animal's horns, and the cochlea of the ear are logarithmic spirals that can be generated using a mathematical constant known as the golden ratio. Mountains and the branching patterns of blood vessels and plants are fractals, a class of shapes that exhibit similar structures at different magnifications. Einstein's $E = mc^2$ defines the fundamental relationship between energy and matter. And a few simple constants—the gravitational constant, Planck's constant, and the speed of light—control the destiny of the universe.

For the last century, physicists have been excited about discovering how reality behaves in terms of mathematical descriptions. This process is akin to discovering some hidden presence in the behavior of the universe—a gnosis. In this sense, aliens and humans would both inherit the tradition of Pythagoras.

The fact that reality can be described or approximated by simple mathematical expressions suggests to me that nature has mathematics at its core, and technological aliens will discover mathematics the same as that of humans. Formulas like $E = \vec{m}c^2$, $\vec{F} = ma$, $1 + e^{i\pi} = 0$, and $\lambda = h/mv$ all boggle the mind with their compactness and profundity.

$E = mc^2$ is Einstein's equation relating energy and mass. $F = ma$ is Newton's second law: Force acting on a body is proportional to its mass and its acceleration. $1 + e^{i\pi} = 0$ is Euler's formula relating three fundamental mathematical terms: e, π, and i. The last equation, $\lambda = h/mv$, is de Broglie's wave equation, indicating matter has both wave and particle characteristics. Here the Greek letter lambda, λ, is the wavelength of the wave-particle, and m is its mass. These examples are not meant to suggest that *all* phenomena, including subatomic phenomena, are described by simple-looking formulas; however, as scientists gain more fundamental understanding, they hope to simplify many of the more unwieldy formulas. I see no reason why aliens will not discover the same truths.

I side with both the the mathematician Martin Gardner and the philosopher Rudolf Carnap, whom I interpret as saying: Nature is almost always describable by simple formulas, not because we have invented mathematics to do so but because of some hidden mathematical aspect of nature itself. For example, Martin Gardner writes in his classic 1985 essay "Order and Surprise":

> If the cosmos were suddenly frozen, and all movement ceased, a survey of its structure would not reveal a random distribution of parts. Simple geometrical patterns, for example, would be found in profusion—from the spirals of galaxies to the hexagonal shapes of snow crystals. Set the clockwork going, and its parts move rhythmically to laws that often can be expressed by equations of surprising simplicity. And there is no logical or *a priori* reason why these things should be so.

Here Gardner suggests that simple mathematics governs nature from the molecular to galactic scales.

Rudolf Carnap, an important twentieth-century philosopher of science, profoundly asserts: "It is indeed a surprising and fortunate fact that nature can be expressed by relatively low-order mathematical functions."

I can easily imagine aliens that worship numbers. In our modern era, God and mathematics are usually placed in totally separate arenas of human thought. But this has not always been the case, and even today many mathematicians find the exploration of mathematics akin to a spiritual journey. The line between religion and mathematics becomes indistinct. In the past, the intertwining of religion and mathematics has produced useful results and spurred new areas of scientific thought. Consider as just one small example numerical calendar systems first developed to keep track of religious rituals. Mathematics in turn has fed back into and affected religion because mathematical reasoning and "proofs" have contributed to the development of theology.

In many ways, the mathematical quest to understand infinity parallels mystical attempts to understand God. Both religion and mathematics attempt to express relationships between humans, the universe, and infinity. Both have arcane symbols and rituals, and impenetrable language. Both exercise the deep recesses of our minds and stimulate our imagination. Mathematicians, like priests, seek "ideal," immutable, nonmaterial truths and then often try to apply these truths in the real world.

Genetic Messages

On one cold December day in 2050, a researcher at the National Biomedical Research Foundation in Washington, D.C., places the corpse of a tarsier (Figure 7.1) inside the sequencing box. A minute later a billion letters representing the animal's genetic sequence are in the computer's memory. The computer analyzes the DNA, finds something peculiar, and notifies the researcher, using its computerized voice. The researcher looks at his computer keyboard, and then he faints, because the computer has determined that a large portion of the tarsier's genetic material codes for the decimal digits of π (3.1415 . . .).

7.1 A tarsier. This tree-dwelling mammal from the East Indies is about the size of a small squirrel.

36

So began one of my science fiction tales in *Mazes for the Mind*. If you were an alien creature trying to code a message using the four symbols of DNA's genetic code (G, C, A, and T), how would you accomplish this, and what message would you encode? I discuss various theoretical methods for doing this in my book *Mazes for the Mind*, but the idea of placing messages in genetic sequences is not entirely fanciful. Joe Davis at MIT once hoped to place messages in the DNA of a bacterium that could duplicate and spread through the galaxy. His collaborator, Dana Boyd, a geneticist from Harvard, has synthesized a short sequence of DNA consisting of 47 base pairs with a brief coded message. When converted to a grid of binary digits, the message appears as a sketch of part of the human body. One hundred million copies of this message have been stored in a vial. At this point Davis and his colleagues do not really plan to disperse these bacterial spores, but Davis has noted that this "may be the only practical way for humans to explore the cosmos."

Could messages from aliens already be enclosed in the genetic material of a harmless virus sent to Earth? The virus could replicate itself, once it had infected an organism. The virus would then spread throughout the populace like an epidemic until some geneticist-cryptographer discovered the message in the genetic sequence and deciphered it.

If I had to design such a message, just as I would do with radio-wave communications, I would put into the message an attention-getting segment not found in nature, followed by instructions on how to read the message, followed by the message itself. To ensure the long-term survival of the attention getter, language lesson, and message, these segments would repeat many times, in case some of the genetic material mutated.

Mark W. Ravera of the Department of Medical Biochemistry at the Rhone-Poulenc Rorer Central Research Center in Collegeville, Pennsylvania, tells me that some institutions have suggested the use of plasmids (short pieces of extrachromosomal DNA) to carry some unique code that could be easily detected at a later date. This imprint would be a short DNA sequence coding for a tiny protein naming the originating institution. For example, plasmids made at the Merck Company would all contain a short DNA stretch that would code for the small protein segment:

Met	Glu	Arg	Cys	Lys
M	E	R	C	K

(The single-letter representation of amino acids employs a standard one-letter naming system commonly used by biochemists today.) This DNA stretch could be detected at any time by a simple genetic technique called DNA hybridization.

The Science Fiction Literature

Although we will have much of our mathematics in common with technological aliens, it's likely that we will never be able to fully understand alien ideas, just as we may never be able to understand the "language" of dolphins or teach a chimpanzee to readily understand our language. The notion of aliens with whom we cannot communicate is explored in Stanislaw Lem's novel *Solaris* (1961), in which an intelligent ocean covers an entire planet. The sentient ocean is studied for years by scientists who are able to recognize the ocean is intelligent, but are totally unable to enter into any kind of meaningful dialogue. Lem insists that we will never truly know the alien ocean and can only resort to models and oversimplifications. Similarly, in Terry Carr's story "The Dance of the Changer and the Three" (1968), seemingly peaceful aliens suddenly murder nearly all the members of a human expedition. In the end, there is no possibility of explaining why the aliens did it. The shocking murder is central to Carr's thesis on the alienness of aliens: "Their reason for wiping out the mining expedition was untranslatable."

How would humanity be affected by a message from the stars? Should we reply? Science fiction writers have often considered the possibility of contact between aliens and humans. In these scenarios, we would have to not only exchange information with aliens, but also determine how to live in harmony with them. In some scenarios, for example in H. G. Wells's *War of the Worlds*, horrible beings invade Earth with the goal of enslaving humanity. In Robert Heinlein's *Starship Troopers*, human military forces fight bugs so different from us in their outlook on life that there is no chance for peaceful coexistence. In Thomas Disch's *The Genocides*, aliens turn the Earth into a huge vegetable patch and eliminate the human pests. In Jack Vance's story "The Gift of Gab," humans try to communicate with intelligent underwater cephalopods who can produce no sounds but use a complex sign language involving movements of their tentacles. In Michael Bishop's novel *Transfiguration*, intelligent beings communicate by rapidly changing the color of their eyes' irises. Their books are discs that quickly flash sequences of colors.

In many science fiction stories, the aliens are already here, but we don't know it. For example, Eric Frank Russell suggests in *Sinister Barrier* that we are being secretly maintained like cattle by invisible vampirelike aliens who feed upon the energy of our negative, violent emotions. Clifford Simak's *Time and Again* and Bob Shaw's *Palace of Eternity* suggest that we already live in symbiotic harmony with invisible beings that provide a secular substitute for souls.

Often in the science fiction literature the first meeting between aliens and humans is filled with distrust. For example, in Murray Leinster's "First Contact" (1945), humans meet aliens, and they both decide to trade ships as the only way to give away equally implicating information about themselves, and they decide to destroy information about the location of their home planets. What would *you* trade with an alien? What gift would you give? Assuming they already have useful information about human civilizations, and you wanted to give them a gift of great value, would you give them a work of art such as a famous Picasso painting? Works of art—such as music, sculptures, and paintings—might be appreciated because they don't require linguistic translation. However, it's not clear that our art would be considered beautiful or profound to aliens. After all, we have a difficult time ourselves determining what good art is.

What do you think aliens would consider beautiful art? Would an alien race of intelligent robots prefer a combination of graffiti-like figures echoing the art of children and primitive societies, or would they prefer the cold regularity of wires in a photograph of a Pentium computer chip? If we were to give these aliens a music tape, they should be able to conclude we have an understanding of patterns, symmetry, and mathematics. They may even admire our sense of beauty and appreciate the gift. What more about us would our art reveal to them? What would alien art reveal to us?

God's Formula

Humans have thought about sending messages to the stars for decades, although there has always been some debate as to what the message should contain. For example, in the 1970s Soviet researchers suggested we send this message:

$$10^2 + 11^2 + 12^2 = 13^2 + 14^2$$

The Soviets called the equation "mind catching." They pointed out that the sums on each side of the "=" sign total 365—the number of days in an Earth year. These imaginative Soviets went further to say that extraterrestrials had actually adjusted the Earth's rotation to bring about this striking equality! Surely it should catch aliens' attention and demonstrate our mathematical prowess.

Personally, I find the Soviet formula arbitrary and not a good candidate for a mathematical expression to send. Rather, I would somehow try to send the most profound and enigmatic formula known to humans:

$$1 + e^i\pi = 0$$

Some believe that this compact formula is surely proof of a Creator. Others have actually called $1 + e^i\pi = 0$ "God's formula." Edward Kasner and James Newman, in *Mathematics and the Imagination,* note, "We can only reproduce the equation and not stop to inquire into its implications. It appeals equally to the mystic, the scientists, the mathematician." This formula of Leonhard Euler (1707–1783) unites the five most important symbols of mathematics: 1, 0, π, e (Euler's number, 2.71828, the base of the natural logarithms), and i (the square root of minus one). This union was regarded as a *mystic union,* almost religious, containing representatives from each branch of the mathematical tree: Arithmetic is represented by 0 and 1, algebra by the symbol i, geometry by π, and analysis by the transcendental e.[3] The Harvard mathematician Benjamin Pierce said about the formula, "That is surely true, it is absolutely paradoxical; we cannot understand it, and we don't know what it means, but we have proved it, and therefore we know it must be the truth."

There is just one problem in sending this formula to the stars. Numbers like π (3.1415 . . .) and e (2.7182 . . .), being transcendentals, contain an infinite number of digits, so we'd have to think of compact ways to represent them. For example, π could be represented diagramatically by indicating the ratio of a circle's circumference to its diameter; e might be represented by an exponential growth curve. However, despite the elegance and profundity of this formula, I admit it is more straightforward to send pulses corresponding to integers such as prime numbers that would easily serve to focus alien's attention on the signal. Occasionally, we, or the aliens, can intersperse instructions on how to build better receivers, transmitters, and so forth.

Whatever the attention-getting signal is, it should be simple. For example, if aliens started with attention-getting signals that were too complex, it would be like trying to make first contact with a prehistoric jungle tribe using a Pentium computer. Instead it would better to start simply by banging on a piece of wood. More sophisticated communication comes later.

The Code Breakers

Imagine how difficult it could be to decipher an alien language. Recall how much difficulty anthropologists and linguists had in deciphering lost languages on our own planet. If Napoleon's troops had not discovered the Rosetta stone near the mouth of the Nile in 1799, Jean Chamollion could not have deciphered hieroglyphics of ancient Egypt. The stone contained an inscription in hieroglyphics, demotic characters, and Greek, and provided a key to the hieroglyphics. Consider also the Etruscan language spoken by the ancient inhabitants of Etruria in Italy, who were early neighbors of the Romans. We still can't decipher it. Etruscan does not seem to be an Indo-European language, and it is known mostly from thousands of short, repetitious inscriptions and an ancient text of 281 lines written on strips of linen cloth. These strips were originally part of a book that was later cut into strips and used in Egypt as a wrapping for a mummy. Another clue is a bronze model of a sheep's liver found at Piacenza that has only 45 words on it.

Despite many attempts at decipherment and some claims of success, the Etruscan records still defy translation. For some words the grammatical category has been established, and for a few of these, a meaning has been assigned.

Even though we know that the Etruscan alphabet was derived from one of the Greek alphabets, and so we can assign sound values to each symbol, we still can't do much translating. If we can't translate Etruscan, could we hope to decipher a message from Alpha Centauri?

If you think Etruscan seems difficult to decipher, there is another fascinating example of how hard it might be to understand an alien message. The mysterious Voynich manuscript here on Earth has defied all attempts to ferret out its meaning. The manuscript, which has been dated at least as early as 1586, is written in a language of which no other example is known to exist. Its script consists of 19 to 28 letters, none of which bears any resemblance to any English or European alphabet. The manuscript is over a hundred pages long, written in running hand.[4] Here is an example:

Of course there would be one major difference between Etruscan and the Voynich language and alien messages beamed to us from outer space. Can you think of it?

Aliens hoping for contact would try to make their messages as *easy* to understand as possible. They would be practicing *anticryptography*, the science of designing codes that are as easy as possible to decipher. Because technological races will understand mathematics, the structure of various atoms and molecules, the positions of stars, or the physics of relativity, this may establish a common starting language.

Some astronomers and mathematicians believe that *symbolic logic* is the best way for intelligent beings from different star systems to communicate. In the 1960s, Hans Freudenthal, professor of mathematics at the University of Utrecht, the Netherlands, attempted to develop a logical language that we could use to communicate with intelligent aliens with whom we have nothing in common. The language is called Lincos, which stands for "lingua cosmica," and it consists of mathematical, biological, and linguistic symbols, including some of those employed by earlier mathematical logicians such as Alfred North Whitehead and Bertrand Russell. In Freudenthal's language, the lexicon and syntax are built up gradually, starting with elementary arithmetical concepts and working up to more advanced abstract ideas. Can you decode any of the message in Figure 7.2 written in Lincos?[5]

Ha Inq *Ha*:
$x \in$ Hom.$\rightarrow$: Ini.xExt$\cdot-$:Ini$\cdot$Cor$x$.Ext $=$
  Cca.Sec 11 × 10^{10111}:
$\vee x \colon x \in$ Bes.$\wedge$: Ini.xExt$\cdot-$:Ini$\cdot$Cor$x$.Ext $>$Sec 0 :
$x \in$ Hom.$\rightarrow \vee$ ⌜y.z⌝$y \smile z \in$ Hom.$\wedge\cdot y = $.Mat $x \wedge \cdot z =$.Pat$x$:
$\vee x \colon x \in$ Bes.$\wedge \vee$ ⌜y.z⌝$y \smile z \in$ Bes.$\wedge\cdot y = $.Mat $x \wedge \cdot z =$.Pat$x$:
$x \in$ Hom.$\rightarrow \colon \wedge t$: Ini$\cdot$Cor$x$.Ext:Ant:$t$Ant:Ini.$x$ Ext
  $\rightarrow$:t Cor$x$.Par$\cdot t$ Cor.Mat x:
$\vee x \colon x \in$ Bes.$\wedge\cdot \wedge t$Etc:
$x \in$ Hom.$\wedge$:$s =$ Ini$\cdot$Cor x.Ext$\cdot$
  $\rightarrow$:$\vee$ ⌜u.v⌝s Cor$u$.Par$\cdot s$ Cor.Mat$x$:
    $\wedge$ PauAnt.sCor v.Par:PauAnt.sCor.Pat$x$
    $\wedge$:s Cor x.Uni$\cdot s$ Cor$u$.s Cor$v$:
$\vee x \colon x \in$ Bes.$\wedge$.Etc:
Hom$=$Hom Fem.$\cup$.Hom Msc:
Hom Fem$\cap$Hom Msc$=$ ⌜ ⌝:
Car:↑xNnc$x$Ext.$\wedge$.$x \in$ Hom Fem
  Pau$>$Car:↑xNnc$x$Ext.$\wedge$.$x \in$ Hom Msc:
$y = Matx$.$\wedge$.$y \in$ Hom$\cdot\rightarrow$.$y \in$ Hom Fem:$\wedge$:
$y = Patx$.$\wedge$.$y \in$ Hom$\cdot\rightarrow$.$y \in$ Hom Msc:
$x \in$ Hom$\cup$Bes:$\rightarrow$:Fin.Cor$x$Pst.Fin$x$#

7.2 *A message to the stars in Lincos.*

Test Yourself

The following are hypothetical mathematical signals beamed to you from aliens. As a playful test, can you decode the significance of any of them? Try holding contests and working in teams. Don't feel bad if none of these messages are comprehensible to you, because far less than 0.0001 percent of the people on our planet could possibly recognize the significance of these alien signals.

Alien Message 1: You are working with a team of scientists who look into the sky and see a flying saucer blinking a light with a sequence of short-duration (0) and long-duration (1) flashes. What could the following patterning signify?

01101010001010001010001000001010000010 0
    01010 . . .

Alien Message 2: A small gray alien hands you a card with the following sequence:

```
2, 71, 828, 1828, 45904, 523536,
0287471, 35266249, 775724709 . . .
```

If your team cannot tell him the significance of this sequence within two hours, the alien will begin examining you with various cold, metallic instruments.

Alien Message 3:   Aliens are waiting for humanity to replace the question mark with the next value in the sequence before they will consider us worthy for further communication:

```
77, 49, 36, 18, ?
```

Alien Message 4:   A message from Alpha Centauri is as follows. What could it mean?

```
14, 15, 92, 65, 35, 89, 79, 32, 38,
46, 26, 43, 38 . . .
```

The President of the United States offers a million-dollar reward to anyone who can solve the mystery.

Alien Message 5:   The following sequences are broadcast using short (0) and long (1) beeps. A Kansas City mill worker who first hears the sounds croaks, "It's the strangest thing ye ever heard. It ain't exactly irregular and it ain't exactly regular, either." The message starts with a short beep and continues to grow ever larger, according to rules the aliens wish us to decipher.

```
                    0
                   0 1
                  0 1 1 0
                 0 1 1 0 1 0 0 1
        0 1 1 0 1 0 0 1 1 0 0 1 0 1 1 0 . . .
```

The solutions to all these messages are given in endnote 6.

Humans have actually transmitted messages consisting of a sequence of 0's and 1's to the Great Cluster in Hercules, using the huge Arecibo radio telescope located in Puerto Rico. Aliens can decode the message by breaking up the digits into 72 consecutive groups of 23 digits each, and arranging the groups one under the other, reading right to left and then top to bottom. If they color the 1's black, they'll be able to discern a schematic representation of a human being toward the bottom of the message. Also encoded in the message are the chemical formulas for elements of the DNA molecule (phosphate group, the deoxyribose sugar, and the organic bases thymidine, adenine, guanine, and cytosine), the numbers 1 to 10 in binary, the atomic numbers for hydrogen, carbon, nitrogen, oxygen, and phosphorus, the number of humans on Earth, and a graph showing the Solar System with Earth displaced towards the figure of the human being. Both the height of the human being and the diameter of the Arecibo telescope are given in units of the wavelength that was used to transmit the message: 5 inches (12.6 cm). How would *you* code a message to be interpreted by extraterrestrials? Should we be sending messages to the stars? What information would you send? Do you think the majority of the people on Earth would be happy to receive an intelligent signal from an advanced extraterrestrial civilization? What effect would this have on politics, religion, and philosophy?

Even if aliens made prodigious attempts to make their messages understandable, how much could we understand from very different creatures transmitting at very long or rapid time scales—for example, a technological species inhabiting the surface of a neutron star, living out their lives in a fraction of a second, as in Robert L. Forward's *Dragon's Egg*. What could we have to say to creatures with whom we have so little in common? As the author John Casti notes in *Paradigms Lost*, perhaps alien science would be no more comprehensible to us than the wiring diagram of an IBM PC is to an aboriginal tribesman. If aliens send political, cultural, and ethical information, the signal may suggest practices or systems that we would find immoral or just plain unworkable, for example the cannibalization of children, the abolition of money, sex with plants, or the rationing of love. This alienness has led John Casti to suggest that the benefits from SETI (the search for extraterrestrial intelligence) are overestimated. Casti believes that even if extraterrestrial intelligences exist, we'll never know them or get any real benefit from them, simply because they are truly and fundamentally alien. I take an opposite view and believe that the mere act of searching is important. Searching and wondering is what science is all about. As Richard Powers has noted: "Science is not about control. It is about cultivating a perpetual condition of wonder in the face of something that forever grows one step richer and subtler than our latest theory about it. It is about reverence, not mastery."

Alien Messengers

You are on a Metro-North commuter train traveling from Croton-Harmon to Ossining, New York, when suddenly the train conductor sees a dark monolith protruding from the tracks. The conductor immediately slams on the train's brakes, bringing the train to a grinding halt inches from the monolith. The conductor exits from the train and approaches the dark shape.

The monolith sits in a capsule of translucent white light as bright as fog on an Autumn morning. The light transforms the monolith into an object of great dignity and unbearable beauty. Within a deep crevice of the monolith are dozens of message-carrying robots resembling ants. Some ants are teachers of the alien language. Others carry knowledge. Others hide out in the underbrush reporting back our actions to the makers of the messengers, a civilization light-years away . . .

I've long been fascinated by the idea that advanced civilizations could be sending robotic emissaries on long interstellar voyages. These message-carrying robots, or "Messengers," can orbit stars and await the possible awakening of civilizations on nearby planets. What would such a Messenger look like? Would it have a head, body, and limbs, or would it look more like the black, stark monolith in *2001: A Space Odyssey*? Would the Messenger robot seem like an intelligent being to us? Could Messengers already be here in our Solar System, hiding in some crater of the moon until we have reached a certain level of technical or moral sophistication?

Because a Messenger may have to wait millions of years before making contact, it would have to be heavily armored to withstand radiation damage and meteorite impacts. It might have powers of self-repair or replication. It could get its energy from sunlight. Ronald N. Bracewell, a leading radio astronomer from Stanford University, talked about such kinds of alien contacts in an article in *Nature*.[7] He believed that Messengers would be more likely than radio contacts for advanced civilizations reaching out to other worlds. Certainly a Messenger could make itself more obvious, or make its signals more easily detectable, than radio signals coming from a seemingly random star.

Consider a Messenger sitting on our Moon as it waits for civilization to evolve. It listens continuously for narrow-band emissions suggestive of a civilization starting to use radio waves. Once it detects the signal, it waits a century for Earth's science to mature, and then it simply sends back to Earth the detected radio signals, producing an echo that might excite our scientists. Perhaps variations in the echo times could actually contain a message from one of these machines.

Bracewell believed that the Messengers might be "sprayed" toward nearby stars by advanced civilizations, and they would not reveal themselves. They would merely report to the aliens when they heard signs of intelligent life, using a star-to-star relay system for efficient communication. If a Messenger were listening to our TV signals today, what would they be transmitting from us to their alien progenitors— *The X-Files, Oprah Winfrey, Seinfeld*, and *The Simpsons?* Would the alien civilization, someday, years in the future, be pondering the utterances of Howard Stern or Rush Limbaugh?

The easiest signals for the Messengers to relay would be our ubiquitous TV and radio shows. In the future, however, with the trend on Earth toward cable television, we might not be leaking such shows into space. Nevertheless, it seems that satellites are used extensively for communications, and airwave signals will always be present.

What might Messengers look like? There is no reason for a Messenger to have a head, body, arms, and legs, even if the aliens had such appendages. Judging from our own experiences with robots, limbs would be too fragile and breakable or could malfunction. More likely the Messenger would be a compact, sturdy shape such as a sphere or icosahedron (a polyhedron with 20 faces).

Another possibility is that the alien civilization intends for us to meet Messengers resembling the aliens to assess how we react to them. In the scenario where aliens were aware of our own appearance, they might construct a Messenger to resemble us so that we would feel more comfortable interacting with it.

I like to imagine the faint possibility that supercivilizations are already linked in a galactic federation of intelligent beings. Perhaps they are experienced in making contacts with emerging intelligence such as our own. If there are superintelligent, technological races in our Galaxy, then the Messengers may already be here in our Solar System, hibernating in wait mode. This is a safe way for the aliens to gain or give information without making the dangerous interstellar voyage. There may be thousands of Messengers swarming in the asteroid belt, reproducing using the large deposits of metals in this region. Their antennas might be pointed at Earth right now—waiting for the next Einstein, Jesus, or Mother Teresa to hit our air waves. . . . Perhaps by monitoring major telephone microwave links between New York and New Jersey, or various communication satellites, aliens could be scanning and downloading the entire contents of the Internet's World Wide Web as they search it for works of art, music, science, and literature. Whether they like it or not, they would also be downloading the ever-increasing pornography, romantic discussions, money-making schemes, Pamela Sue Anderson photos, conspiracy theories, and all manner of the wild and the weird.

Other Contact Methods

There are two types of alien signals we might someday receive: those that are accidentally leaked from an alien civilization's private radio transmissions (like commercial radio and TV broadcasts and military radar waves) and those that are beamed to us intentionally. It would be quite difficult to detect the leakage radiation because it weakens rapidly with distance and would probably require radio telescopes larger than any in use now on Earth. If aliens were *deliberately* signaling Earth, we'd have a better chance of detecting the message. Radio waves would be the obvious medium because they can travel interstellar distances with considerably less interference than signals at other wavelengths, such as visible or infrared light. Alien messages could be encoded in carrier waves, modulated signals like those of AM or FM transmissions, or in pulses that would be detectable at greater distances, given the same amount of energy. In addition, the aliens could increase the odds of detection by transmitting at frequencies around 1420 MHz (megahertz), where interstellar gas, dust, and Earth's atmosphere cause little interference.

Communication methods other than radio are possible. For example, some scientists feel that it is only a historical accident that lasers weren't discovered before radio as a means for long-range communication. Is it possible that another civilization has become very sophisticated in the use of lasers? Could we even detect a laser beam sent to us from another solar system?

Here is a recipe for success with laser signals. If aliens passed a laser beam through an optical system with a 200-inch (513 cm) reflector (comparable to the one in the telescope on Mount Palomar in California), their optical system would focus a 200-inch-wide beam. Under favorable viewing conditions, a person on Earth could view the laser beam 0.1 light-years away with the unaided eye. With the proper telescopes, laser technology could send observable beams between nearby star systems. If the aliens wanted to make sure we could see their signals, they should use appropriate wavelengths of light to make sure the laser signal is not hidden by the brilliance of their own nearby sun. We might detect their laser light by scanning the light of nearby stars for strong intensities at unusual wavelengths. Today our own laser signaling capability is not as potent as our radio waves. Currently we can send detectable *radio* messages out 1,000 light-years, bringing some 2 million stars within our range, whereas laser signals can only travel about 10 light-years with our current level of technology.

To provide evidence of their existence, more advanced civilizations might even cloak their suns in clouds of material that absorb some unusual wavelength of light. As a result, the entire galaxy could view this magnificent beacon of intelligence. The astrophysicist Frank Drake suggests that a supercivilization could place a technetium cloud around their star. This radioactive metal is observed on Earth only when it is produced artificially, and only weakly on the sun, because it is short-lived and rapidly decays away. Drake estimated that aliens could mark a star using only a few hundred tons of light-absorbing substance spread around the star.

I like to stretch my mind to consider other possible, far-out achievements of highly advanced civilizations. Perhaps someday advanced aliens will be able to beam us radio information allowing us to assemble copies of the aliens. These copies could be approximate or simplified in ways that would permit transmission. As our science and communication methods improve over the next century, even we might be able to transmit the information needed to encode the essential biochemical components of an egg cell. Aliens in turn should be able to teach us how to build their versions of cars, cows, and sunflowers.

SETI Awaits Alien Broadcasts

The world's first search for extraterrestrial intelligence, or SETI, was conducted in 1960 by Frank Drake, using an 85-foot parabolic radio antenna located in Green Bank, West Virginia. This early project, called Ozma after a book in L. Frank Baum's *Wizard of Oz* series, searched for alien signals emanating from two nearby stars, Epsilon Eridani and Tau Ceti.

Frank Drake found nothing, and to this day no definite extraterrestrial signals have been recorded by numerous radio searches. Two SETI searches have produced tantalizing hints, however, and astronomers are trying harder than ever to find an unambiguous signal.

Back in the sixties, Drake had come to believe that the conditions for life were common in the universe, that sunlike stars with water-bearing planets might pepper our Galaxy. Recent discoveries have confirmed many of Drake's assumptions. For example, astronomers have discovered several planets that orbit other stars and have found complex organic molecules floating in interstellar space.

The recent interest in possible Martian fossils has increased interest in SETI radio searches for alien civilizations, according to Drake, who is now president of the SETI Institute in Mountain View, California. He's had to listen to two U.S. senators ridicule his efforts over the years. The Harvard University zoologist Ernst Mayr, an outspoken critic of SETI, argues that the huge evolutionary odds stacked against intelligent life make SETI a waste of time. However, in spite of the dissent, interest in SETI continues with renewed vigor, on the part of both scientists and movie producers.

In the movie *Contact*, based on Carl Sagan's book of the same name, Ellie Arroway, played by Jodie Foster, detects an alien broadcast, using sensitive radio receivers on Earth. As a matter of fact, as in the movie, a female Ph.D. astronomer, Jill Tarter, directs Project Phoenix, a SETI program based in Mountain View, California. Project Phoenix is currently the most sensitive search for extraterrestrial intelligence and the only one capable of detecting pulsed signals. The privately funded program continued NASA's search after Congress terminated its support in 1993. Phoenix uses some of the largest radio telescopes on Earth and points them at more than 1,000 stars within 150 light-years. The program has developed special electronic equipment that scientists carry in a trailer to different telescopes. For example, the Phoenix electronic components have been hooked up to the 140-foot (43 m) antenna at Green Bank, near the smaller dish used by project Ozma, and at the giant 1,000-foot (305 m) dish at Arecibo, Puerto Rico. The Southern Hemisphere phase of Project Phoenix uses the 210-foot (64 m) Australian telescope in Parkes, New South Wales, to look at 202 sunlike stars.

Incidentally, in the movie *Contact*, Ellie detects the alien radio signal while listening with earphones. In reality, Project Phoenix examines 28 million channels simultaneously, which would require lots of earphones! Instead, computers scan for signals and only alert the astronomers when interesting ones appear.

Where Phoenix listens to one star at a time, other SETI searches sweep the whole sky, looking for powerful but intermittent signals that may drift in frequency, or "chirp." These sky surveys are led by Project SERENDIP, sponsored by the University of California, Berkeley. It has a present capacity of 4 million channels and will soon advance to 68 million channels, using an improved software package called SERENDIP IV to scan the sky.

The third major SETI project is Harvard University's BETA, which uses an 85-foot (26 m) dish located in the Massachusetts town of Harvard. Although its smaller antenna affords less sensitivity than those of Phoenix and SERENDIP, BETA's computer can scan more than 2 billion frequency channels every 16 seconds. It lacks the backup dish that Phoenix employs to screen out radio-frequency interference, but its single dish checks three parallel beams to filter out Earth-based noise.

BETA improves upon its predecessor, META, which was funded in 1985 by the creator of *E.T.*, Steven Spielberg. META recorded 37 radio signals that appeared to be of extraterrestrial origin but were never repeated. These signals were reminiscent of a strong signal from the constellation Sagittarius detected by scientists at Ohio State University in 1977. That signal, however, also disappeared before scientists could recheck it. Most of these promising signals originate along the galactic plane, the flat region of our Galaxy, where stars, planets, and possible life sites are concentrated. However, only a signal that persists long enough to allow independent confirmation by other radio telescopes would have a chance of convincing the scientific community.

Most SETI researchers remain optimistic because their detection capability increases dramatically every year. Proposals for twenty-first-century SETI projects are already on the table. For example, the square kilometer array interferometer, SKAI, would assemble a large observing area from an array of smaller dishes. The array's sensitivity would allow detection of the kind of leakage from outer space that Earth broadcasts *to* space every day. An ideal telescope would be the size of Arecibo and located on the side of the Moon facing away from Earth. Scientists using such a telescope would not have to worry about separating a possible alien signal from all the radio noise produced on Earth.

Aside from the three big SETI searches, there are impressive smaller searches for extraterrestrial intelligence. The SETI League is a worldwide group of amateur and professional radio astronomers, radio amateurs, microwave experimenters, and digital signal–processing enthusiasts who have banded together in a systematic, scientific search of the heavens to detect evidence of aliens (Figure 7.3). There are grounds for thinking that people

7.3 SETI League executive director Dr. H. Paul Shuch with a portable radio telescope. This system serves as a test bed for the hardware and software to be used for the Project Argus all-sky survey. The actual antennas used for SETI are much larger. (SETI League photo by Muriel Hykes, used by permission.)

7.4 Dr. H. Paul Shuch poses in front of one of the antennas of the National Radio Astronomy Observatory's Very Large Array, Socorro, New Mexico. The VLA consists of 27 of these 82-foot-diameter (25 m) dishes, each weighing 230 tons. (SETI League photo, used by permission.)

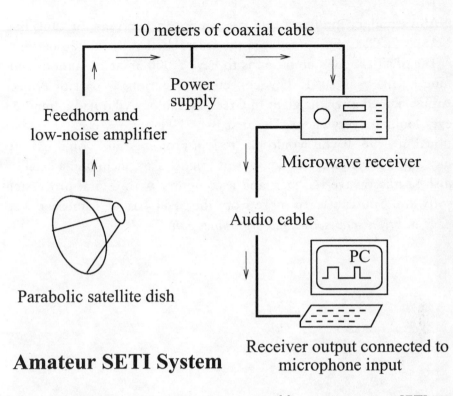

10 meters of coaxial cable

Power supply

Feedhorn and low-noise amplifier

Microwave receiver

Audio cable

PC

Parabolic satellite dish

Receiver output connected to microphone input

Amateur SETI System

7.5 Build your own SETI station. Diagrammed here is an amateur SETI system that costs about $7,000. If you use an existing computer and a surplus dish, and build some hardware from kits, the cost can be as little as $1,000. The price of an "interplanetary call" is continually decreasing!

working from their backyards can make a real difference in the hunt for intelligent signals from space. For a start, the kind of sensitive equipment that professionals were using for SETI some twenty years ago is now available off the shelf, thanks to the revolution in satellite TV. Also, by working together in large numbers, amateurs can monitor the entire sky, something the professionals with their big SETI dishes would find difficult. A big dish sees only about a millionth of the sky, so it can easily miss an alien signal (see Figure 7.4).

The SETI League, founded in 1994 as a membership-supported educational and scientific organization, was also established in response to Congress's terminating all NASA funding for SETI in 1993. Before its funding was cut, NASA's project consumed one tenth of 1 percent of

NASA's budget, or five cents per American per year. In canceling NASA's SETI, Congress reduced the federal deficit by 0.0006 percent.

The SETI League's ambition is to have 5,000 amateur stations running by the year 2001. This collection of stations is part of Project Argus, named after the giant in Greek mythology who had a hundred eyes looking everywhere. The first five SETI stations became operational in 1996. If you would like to join Project Argus, you'll have to buy a few basic pieces of equipment (Figure 7.5), including a satellite dish, a microwave receiver, and a computer with digital-processing software. You can learn more from the SETI League's Internet Web page at *http://seti1.setileague.org/homepg.htm*.

ALiEN TRAVEL

If they are not an advanced race from the future, are we dealing instead with a parallel universe, another dimension where there are other human races living, and where we may go at our expense, never to return to the present? From that mysterious universe, are higher beings projecting objects that can materialize and dematerialize at will? Are UFOs "windows" rather than "objects"?

—Jacques Vallee

We have never denied that it is possible, indeed probable, that other forms of life, even intelligent life, exist in the universe. But this is different from the belief that we are now being visited by extraterrestrial beings in spacecraft, that they are abducting people, and the there is a vast government cover-up.

—Paul Kurtz

High-Speed Travel

If aliens wished to visit us from far-away stars using standard space-travel methods, their journeys could take very long indeed, even if they traveled on high-speed spaceships. It seems that the laws of physics would prevent them from traveling faster than the speed of light.

The speed of light constrains the motions of any objects with mass. To better understand this constraint, consider what would happen if you lived in a universe where the speed of light were only 60 miles (97 km) an hour. You'd be quite frustrated, because as your foot pressed your car's gas pedal, you would never quite reach the speed limit, no matter how hard you pressed the pedal. Now imagine you are on a high-speed rocket that faces the same problem as it approaches c, the speed of light in a vacuum. According to Einstein's special theory of relativity, an increase in an object's speed also increases its mass, and the mass becomes infinite at the speed of light.[1] To move an infinite mass would require infinite energy. This suggests that as a spaceship nears light speed, its mass increases toward infinity—meaning that to accelerate all the way to the speed of light, an infinite amount of fuel is needed. Einstein's special theory of relativity also suggests that by traveling closer to velocity c you can make your journey seem as short as you like—though only to yourself! This is because clocks on the spaceship run more slowly than stationary clocks on Earth, making your long journey appear to you to take a few hours. To those on Earth, your journey may appear to take centuries.

Actual application of high-speed travel would be difficult because of a very high rate of collision with atoms in outer space. For example, there is about one hydrogen atom in each cubic centimeter of space, and if many collide with your spaceship there would be an intense lethal radiation in the form of gamma rays, because the collisions promote nuclear fusion reactions that produce gamma radiation.

Tachyonic Aliens

As we just discussed, the popular wisdom is that aliens could not travel faster than the speed of light. However, in my book *Time, A Traveler's Guide*, I discuss the strange possibility of tachyonic aliens that can travel faster than light. As background, Albert Einstein's theory of relativity doesn't preclude objects from going faster than light speed; rather, it says that nothing traveling slower than the speed of light (for example, you and me) can ever travel faster than 186,000 miles (299,000 km) per second, the speed of light in a vacuum. However, faster-than-light (FTL) objects may exist, so long as they have never traveled slower than light. Using this framework of thought, we might place all things in the universe into three classes: those always traveling less than 186,000 miles per second, those traveling exactly at 186,000 miles per second (photons), and those always traveling faster than 186,000 miles a second.

In 1967, the American physicist Gerald Feinberg coined the word *tachyon* for such hypothetical FTL particles. The name comes from the Greek word *tachys*, "fast." In contrast, *tardyons* are the slower-than-light particles with which we are familiar (for example, protons and electrons). Sometimes tardyons are known as *ittyons,* from the Hebrew for "slow." Tardyons have mass, but the very light ones are relatively easy to accelerate to near light speed. For example, the electrons that produce an image on a TV screen travel at about 30 percent of Einstein's limit when they hit the phosphor screen. Electrons in the linear accelerator at Stanford University can be made to lag behind light speeds by only a few parts per billion, less than 1 mile (1.6 km) per hour.

Aside from the tardyons, there are also the massless *luxons,* which travel *only* at the velocity of light. Luxons include photons, hypothetical gravitons, and possibly the neutrino. Television, radio, and radar waves are just lower-frequency versions of visible light and therefore also travel at light speed, as does electromagnetic radiation of higher frequencies such as ultraviolet light, X-rays, and gamma radiation.

We know for sure that as a particle's speed increases, it becomes heavier and more resistant to further acceleration. In laboratories all over the world, the increase in mass of elementary particles as they approach the light barrier is well known. However, physicists can imagine bypassing the mass-increase barrier not by accelerating a particle to light speed and beyond, but by putting enough energy together in one place to create a particle, like a tachyon, that is *born* traveling faster than light.

If an alien made entirely of tachyons came toward you from his spaceship, you would see him arrive at your doorstep first, and then see him leave his ship. This seems strange, but the image of him leaving his ship would take longer to reach you than his actual FTL body. Stranger yet, the creature (see Figure 8.1) would visually appear to be traveling away from you, back to his ship. In short, you would see a tachyonic alien as receding backward in time! This is not some optical trick—it is an actual fact predicated on the theory of relativity. These ideas are usually discarded in science fiction stories. For example, if Captain Kirk in *Star Trek* were looking out the *Enterprise*'s window in search of a ship coming toward the *Enterprise* at warp (FTL) speed, he wouldn't actually see the object until after it had arrived! Captain

Kirk's universe would be filled with ghost images of spaceships that long ago arrived where they were going at warp speed.

If a tachyonic Robin Hood shot an arrow at an apple, we would see the hole in the apple before we saw Robin Hood pull back on his bow. In fact, once we had information about the hole in the apple, we might be able to send a tachyonic message to Robin Hood forbidding him to shoot, thus creating a time paradox.

Science fiction authors have often used tachyons for achieving FTL travel or for information exchange. For example, in Bob Shaw's *The Palace of Eternity*, a million-ton tachyonic spaceship travels at 30,000 times the speed of light! In Gregory Benford's novel *Timescape*, future humans use tachyonic messages to warn the past about ways to avoid severe ecological damage to Earth. Even though we can't convert tardyons to tachyons, it might be possible to generate tachyons, send

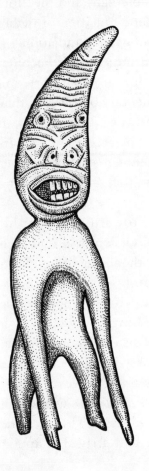

8.1 A tachyonic alien. If an alien, made entirely of tachyons, came toward you from her space-ship, you would see her arrive and then see her leave her ship. Her image leaving her ship would take longer to reach you than her actual FTL (faster-than-light) body. Oddly, the creature would visually appear to be traveling away from you. This means you would see a tachyonic alien as receding backward in time!

them out, and modulate them to transmit information. This would be a means for sending information to the past.

Beings Made of Pure Energy

The change of an object's mass as a function of its speed was first experimentally observed by the German scientist Walter Kaufmann (1871–1947). In particular, Kaufman first observed this mass increase for electrons, the lightest particles known, aside from massless particles (luxons) such as the photon and the hypothetical graviton, or the neutrino, which might have a very slight mass. It turns out that if an electron (which at rest is 2,000 times lighter than a proton) is accelerated to near light speed, the electron is measured to carry a momentum equivalent to a proton. If an electron were accelerated to 0.99999999999999999999999999999999999999999999999999 99999999999 times the speed of light, this tiny particle would slam into you with the impact of a Mack truck traveling at 60 miles (97 km) per hour. This is why it would probably be impossible to accelerate a spaceship such as the *Enterprise* on *Star Trek* close to light speed and beyond. As Lawrence Krauss notes in *The Physics of Star Trek*, "All the energy in the universe would not be sufficient to allow us to push even a speck of dust, much less a starship, past this ultimate speed limit of light." Because of this mass-variation effect, even when the *Enterprise* is using its "impulse drive" powered by nuclear fusion, each time the *Enterprise* accelerates to half the speed of light, it would have to burn 81 times its entire mass in hydrogen fuel.

I should also point out that not just light but all massless radiation must travel at the speed of light. This means that many beings made of "pure energy"—commonly found in science fiction—must travel at the speed of light. They'd have trouble slowing down, and their clocks would be infinitely slower compared to our own!

Despite these difficulties, we may someday be able to travel great distances at sub–light speed using futuristic energy sources making use of matter/antimatter reactions,[2] or zero point energy—the imperceptible yet titanic legacy of the Big Bang. Although these methods seem very far-fetched today, the late Nobel Prize–winning physicist Richard Feynman one remarked that there is enough energy in a cubic meter of space—any space, anywhere—to boil all the oceans of the world. If this can be tapped—and, according to the futurist and science fiction author Arthur C. Clarke, there is evidence that this could already be happening in some laboratories—travel to the planets, and even the stars, will become less expensive and easier.

What Is Death?

Aliens *could* visit us if they were willing to take a long slow journey, perhaps hibernating along the way or having many offspring during the trip. Generations would be born and die on the ship before the journey was complete. If the aliens had long life spans, or used suspended animation or freezing to prolong their lives, such a journey would be possible. Although it is hard to predict alien psychology, no doubt many aliens would find 10,000-year trips unappealing. However, even at speeds one hundredth the speed of light, aliens could conceivably colonize the entire galaxy within 5 million years—a short time relative to the history of Earth. If there were aliens that had adapted to living freely in the vacuum of space, they certainly would spread and speciate the galaxy.

Aliens may have perfected the process of *cryonics*—freezing a body to be revived at some later time—in order to make long space journeys possible. Humans have been experimenting with cryonics since the fifties, when hamster brains were partially frozen and revived by the British researcher Audrey Smith. In the 1950s and 1960s, golden hamsters were cooled until 60 percent of their brain water had been converted to ice and then made a complete recovery with no behavioral abnormalities.[3] If hamster brains can function after being frozen, why can't aliens achieve perfection in their cryonics? In the 1960s, the Japanese researcher Isamu Suda froze cat brains for a month and then thawed them. Some brain activity persisted. At about the same time, isolated cat brains were also treated with 15 percent glycerol and cooled to –4 degrees F (–20 degrees C) for five days. When warmed they returned to normal brain function as determined by EEG measurements.[4] Aliens could certainly freeze entire embryos if their science were similar to our own, because thousands of healthy human babies have been born from frozen embryos.

Evidence for the possibility of biological resurrection continues to grow. Living animals can be "stopped" by placing them in suspended animation by freezing or by dehydration. Later, the animals can be revived at the scientist's leisure. For example, tardigrades, members of the phylum Tardigrada—small animals with four pairs of short, stubby legs armed with claws—can be dried out and then rehydrated like a backpacker's dinner. The creatures return from life after apparent death.

Alien tardigrades could travel through space, like spores, inside meteorites or comet ice. Tardigrades can withstand extremely low temperatures and desiccation. Specimens kept for eight days in a vacuum, transferred for three days into helium gas at room temperature, and then exposed for several hours to a temperature of −458 degrees F (−272 degrees C) came to life again when they were brought to normal room temperature! Sixty percent of specimens that had been kept for 21 months in liquid nitrogen at a temperature of −310 degrees F (−190 degrees C) also revived.

Artemia franciscana, also known as sea monkeys or brine shrimp, can live for years without oxygen. In 1997, James Clegg, a biochemist at the University of California at Davis, took dried-up brine shrimp embryos and rehydrated them in oxygen-free water at room temperature. Without oxygen, the embryos looked like floating lifeless corpses. Four years later, when Clegg exposed the embryos to oxygen, 60 percent resumed normal development. Clegg suspects that a class of proteins called molecular chaperones protect the shrimps' proteins from degrading, allowing the shrimps to return from the dead. If the water were colder, the shrimp might last for decades.

Other creatures can survive for years in suspended states. Viable bacterial endospores over 9,000 years old have been isolated from sediment collected from lake bottoms, but such abilities are not restricted to microscopic critters. Resting-stage eggs of copepods (small crustaceans) that have remained buried in sediments at the bottom of freshwater lakes for over 400 years can still hatch.

Another interesting example of "suspended animation" in a complex organism occurs in the desert snail. In 1846, two specimens were presented to the British Museum (now the Natural History Museum) in London to be displayed as *dead* exhibits. Four years later, in March 1850, someone found that one of the snails was still alive. This hardy little creature lived for another two years before it fell into a torpor and died.

If aliens naturally required lengthy periods of sleep, perhaps the idea of hibernation during long voyages in space would be more appealing to them. However, it is difficult to imagine what it would be like to meet an alien who required far more sleep than we do. Efficient interactions between humans and these alien *narcophiles* (sleep lovers) could be difficult if they spent almost all their time sleeping. This idea is not very fanciful when we consider animals such as the armadillo, opossum, and sloth, which spend up to 80 percent of their lives sleeping or dozing. Koala bears average 22 hours of sleep a day.

At the other end of the scale is the frenetic shrew, which has to eat constantly or starve to death. It literally has no time for sleep. The swift actually sleeps while flying, turning off half of its brain for two hours while the other half pilots the bird. Then the sleeping hemisphere wakes up and gives the active hemisphere a rest.

It is interesting to speculate on whether narcophilic aliens could ever really develop advanced technology. From an evolutionary standpoint, I'd expect their ancestors to have spent most of their time hiding from predators, for they would be vulnerable when asleep, and it may be difficult for such a species to evolve into creatures actively engaged in exploring the stars or in the creation of technology.

Could an alien life span be many times longer than ours? To aliens we might appear as dying butterflies rising and falling in a comparative instant. What seems impossible for us because of time, organization, or resource constraints may be easy for long-lived aliens existing outside the maelstrom of time. Thus, although they might not approach the speed of light, it may be of little concern to immortal beings unaffected by the gentle acid of time as they estivate in the deserts of vast eternity.

Finally, there is always the possibility that alien civilizations are able to make use of cosmic worm holes in outer space to jump between remote areas of the universe or to other universes. Perhaps the Starship *Enterprise* could zoom around the universe if the fabric of space-time in front of it were crumpled while that behind it were simultaneously stretched—in effect, bringing two far-away locations close together. Aliens may also be able to travel between parallel universes. All of these mathematical and physical ideas are discussed in my book *Black Holes—A Traveler's Guide*. For now, such theories are rather speculative.

My favorite science fiction aliens that travel between universes are "Species 8472" from the TV show *Star Trek: Voyager*. These powerful and terrifying beings come from an entirely different universe where there are very few other life-forms. Their universe contains no stars or planets but instead is filled with an organic fluid. The creatures are said to travel to our universe through "quantum singularities," using ships that appear to be life-forms complete with vertebrae.

Members of Species 8472 are coldly malevolent. Their only desire is to conquer, and there is no negotiating with them. Like the Borg ideology of assimilation and "resistance is futile," Species 8472's ideology is "The weak will perish."

8.2 Species 8472 from Star Trek: Voyager. *(Drawing by Brian Mansfield.)*

What Would They Do to Us?

If aliens *could* travel to Earth, once they arrived what would they do to us? In the early 1930s, the Russian author Konstantin Eduardovich Tsiolkovsky believed that a superior, godlike race would painlessly eliminate animals on other worlds rather than see them endure the needless sufferings of evolution and struggle for existence. According to Tsiolkovksy's reasoning, such beings would be like good gardeners, weeding out lower animal species, harmful bacteria, and valueless plants, except for a few specimens they would keep as laboratory samples.

Do you think such godlike creatures would have a profound respect for the diversity of life, or would they have no more regard for us than we would for ants? What if the first aliens with whom we come in contact are only slightly superior to us? Perhaps the mature adults would be wise and kind, but the alien teenagers, out of boredom, might get their kicks in strange ways as they explored a human society of vastly inferior creatures. Children on Earth often experiment with frogs and worms, or pour salt on slugs, to view the results.

Some scientists have argued that if aliens were hostile and warlike, they may have destroyed themselves with their own weapons long before they could attempt interstellar travel. Despite imaginative movies such as *Mars Attacks* and *Independence Day*, it is doubtful that aliens would want to enslave us or use us for food. Any aliens sufficiently advanced to traverse interstellar distances would likely have great stores of energy.

However, given our own history, I'm not confident that a space-faring society capable of crossing interstellar distances must consist of benign or caring beings. The ocean voyages of Europeans during the Renaissance period of exploration were in some ways comparable to today's space explorations. One of the motivations for the exploration of the New World was to convert the inhabitants to Christianity, by force if necessary. In fact, during this period of exploration, the Native Americans were not useful for any specific task in the courts of Spain and France, yet they were shipped there as objects of curiosity and for prestige purposes. Might aliens wish to take us back home as showpieces?

There are many stories of first contact in our own history. For example, the strong, warlike Aztecs in Mexico reacted to the Spaniards with adoration. Because the Aztecs had myths foretelling the coming of a white god bringing fancy gifts, the they were more easily conquered than they might have been had they recognized the Europeans as the threat that they were. Imagine what would happen if warlike aliens came to Earth looking like stereotypical images of Jesus? On the other hand, imagine the consequences of a wise and kind alien race coming to Earth in the form of devils, hyenas, rattlesnakes, or skeletons. As much as we'd try to downplay our prejudices, the appearance of aliens would shape our interactions.

Do you think aliens would stage a militaristic invasion of Earth using fighter craft, as in the movie *Independence Day*? Although I believe that aliens could be hostile, the arrival of a fleet of warships for a *militaristic* takeover of the Earth is unlikely. It would be easier and safer for them to threaten world governments by redirecting a large asteroid or several thousand projectiles so that the objects would collide with Earth. They could use cobalt-bomb clusters in which each cobalt bomb is an ordinary atomic bomb encased in a jacket of cobalt. When a cobalt bomb explodes, it spreads a huge amount of radiation. If enough of these bombs were exploded, humans would perish. A few pounds of poison produced by the botulis bacteria is sufficient to kill all human life. If the aliens want to destroy us while preserving the planetary ecology, weather, and terrain, they could drop an *engineered* virus that spreads through casual contact. If they wanted to exploit us, they could work human agents into powerful government and industry positions. This is closer to an *X-Files*–like conspiracy and also reminiscent of John Carpenter's movie *They Live*, in which aliens treat Earth like a Third World country.

Why No Visitors?

Predicting alien technology is very difficult. Advanced aliens may be able to use exotic fuels or worm holes in space connecting different regions of the universe, but there still seems to be incredible dangers. If very high speeds are involved, hitting a tiny meteoroid would be like running into a hydrogen bomb. Cosmic rays in outer space would also pose a danger because of the threat of radiation sickness and leukemia—unless aliens had biologies that could easily repair radiation damage or they could build their craft with very thick, heavy walls.

If we never encounter visitors from another world, it could mean that space-faring life is extremely rare, interstellar flight is extremely difficult (or judged not to be worth the effort), or technological civilizations destroy themselves before embarking on such an arduous task. But perhaps there is another possibility—maybe there *are* signs of alien life all around us that we have not looked for or have not understood. Imagine that our civilization is quarantined by a galactic cartel as a kind of zoo, not to be touched, only to be observed, either because aliens don't want to contaminate our world with alien ideas or be contaminated by us. They could have no desire to interfere with us any more than we want to go out and buy a net to catch butterflies or seahorses. In 1973, John A. Ball, a radio astronomer at the Harvard-Smithsonian Center for Astrophysics, proposed the zoo hypothesis in *Icarus*, an international journal of Solar System studies. He wrote, "The perfect zoo (or wilderness area or sanctuary) would be one in which the fauna do not interact with, and are unaware, of their zoo-keepers."

If an alien society had conquered all their problems—violence, population, pollution—and spent their days percolating in a tai chi–like glow of inner peace, they might not have the slightest desire to reach out to other worlds. What would be their state of mind? Would their intellectual curiosity dissipate into the wisps and eddies of wind? Or would they become so bored with their own world that they would search out novelty in other cultures. Imagine the excitement *we* would have in learning about alien novels, art histories, and religious philosophies.

Intelligence seems to begin with a simple awareness of the environment, and it proceeds toward awareness of the self and then to abstract reasoning. Curiosity is probably universal. From my experience with animals on Earth, I believe curiosity is an essential part of intelligence. I even see all the hallmarks of extreme curiosity in my family's pet guinea pig. In my opinion, intelligent aliens will also be curious aliens.

ALIEN ABDUCtION

With temporal lobe epilepsy, I see things slightly different than before. I have visions and images that normal people don't have. Some of my seizures are like entering another dimension, the closest to religious or spiritual feelings I've ever had. Epilepsy has given me a rare vision and insight into myself, and sometimes beyond myself, and it has played to my creative side. Without temporal lobe epilepsy, I would not have begun to sculpt.

—A female temporal lobe epileptic

When we explore phenomena that exist at the margins of accepted reality, old words become imprecise or must be given new meanings. Terms like "abduction," "alien," "happening" and even "reality" itself, need redefinition lest subtle distinctions be lost.

—John Mack

I am not a temporal lobe epileptic.

—Whitley Strieber

What Is Alien Abduction?

In my book *The Alien IQ Test*, I mention my fascination with the recent frenzy of interest in alien abduction. In the classic UFO abduction scenarios, abductees tell us that they experience bizarre dreams, memory flashes, or even physical symptoms weeks after an alien encounter. Some "abductees" undergo hypnosis in an attempt to recover memories of events that occurred during unexplained lapses of time. These lapses are sometimes called "missing time," when events that seemed to be of a few minute's duration actually took hours. Under hypnosis, abductees reveal that they have been led, sometimes floated, into disk-shaped craft by aliens with large heads and large, slanted eyes (Figures 9.1, 9.2, and 9.3). In these ships, abductees undergo a medical examination of some kind. Some believe that the abduction phenomenon represents evidence for other dimensions beyond space-time, and that aliens may not come from ordinary space but from a "multiverse" all around us. In my opinion, there is not sufficient hard evidence that alien abduction exists outside the mind of the abductee.

9.1 One kind of alien typically reported during abduction experiences. (Drawing by Carol Ann Rodriguez.)

9.2 Another kind of alien
typically reported during
abduction experiences.
(Drawing by Brian
Mansfield.)

9.3 Another alien variation typically
reported during abduction experiences.
(Drawing by Michelle Sullivan.)

About a year ago, as part of my research into alien abduction phenomena for a chapter in my book *Strange Brains and Genius*, I surrounded myself with an array of books and articles: Whitley Strieber's *Communion* and *Transformation*, John Mack's *Abduction: Human Encounters with Aliens* (revised edition), Susan Blackmore's "Alien Abduction," Philip Klass's *UFO Abductions*, Budd Hopkins's *Intruders*, Eve LaPlante's *Siezed*, and C. D. B. Bryan's *Close Encounters of the Fourth Kind: Alien Abduction, UFOS, and the Conference at M.I.T.*

After reading from about 10 o'clock to 11:30 at night, I went to sleep and had a very vivid alien encounter myself. Two beings floated through the screen covering the open window. When standing on the carpet, their heads were just slightly higher than the windowsill. I noticed a slight odor of limes, or perhaps some fragrant flower. The two creatures were hairless and had no ears. Their huge black eyes had no whites or pupils—just like the descriptions in the book I had been reading.

Do I really think two gray beings from another dimension came into my bedroom that night? They were probably a fragment of a realistic dream inspired by my reading of *Communion* and other alien abduction books just before sleeping. In the morning, when I checked the carpet beneath the bedroom window, there were no signs of strange footprints or debris from the bushes. The screen in the window was in place.

Although I keep an open mind, I still await "proof" that alien encounters are not simply vivid mental experiences without physical reality. On the other hand, there does seem to be evidence that such experiences can be products of our minds. For example, patients with temporal lobe epilepsy (TLE) often believe that they are *controlled* from the outside, either by God or by alien creatures from outer space. (The temporal lobe is located in the bottom middle of the brain.) The TLE expert Eve LaPlante, in *Seized,* suggests that the best-selling author Whitley Strieber has TLE. In 1987 Strieber wrote the book *Communion,* which described his abduction by short aliens with two dark holes for eyes. Strieber exhibits various symptoms of TLE: *jamais vu,*[1] formication,[2] vivid smells, hallucinations, rapid heartbeats, the sensation of rising and falling, and partial amnesia. Magnetic resonance imaging of Strieber's brain reveals "occasional punctate foci of high signal intensity" in his left temporoparietal region, which is suggestive of scarring that could lead to TLE. According to skeptics, Strieber incorporated his occasional memory lapses and periods of altered consciousness into his book *Communion,* for which he received a $1 million advance from his publisher.

TLE probably accounts for some reported out-of-body UFO abduction experiences. In fact, a significant number of abductees feel mild epilepsylike symptoms in advance of an "abduction." For example, some abductees feel heat on one side of their faces, hear a ringing in their ears, and see flashes of light prior to an abduction. Others report a cessation of sound and feeling, or an overwhelming feeling of apprehension. Such sensations are often reported by epileptics.

In my opinion, alien abduction stories tell us about the workings of the mind. Michael Persinger, a neuroscientist at Laurentian University in Sudbury, Ontario, found that people with frequent bursts of electrical activity in their temporal lobes report sensations of flying, floating, or leaving the body, as well as other mystical experiences. By applying magnetic fields to the brain, he can induce odd mental experiences—possibly by causing bursts of firing in the temporal lobes. For example, he has made people feel as if two alien hands grabbed their shoulders and distorted their legs when he applied magnetic fields to their brains. However, some psychiatrists, such as John Mack, do not accept the TLE theory of abduction because no one has proven that many abductees have excessive bursts of electrical activity in the brain's temporal lobes.

LaPlante and other TLE experts suspect that our modern fascination with ESP, out-of-body travel, past-life regression, and other paranormal phenomena may be the result of mild, undiagnosed TLE. Strieber, who has received many letters from people who report having similar UFO abduction experiences, notes:

> It's a terribly important and fundamental human experience—perceptions that come from the level of mind that isn't interrupted by the rational structures that animate most of our thought. It's a kind of memory, a form of perception, or a mechanism of consciousness, something inexplicable that the mind attaches an explanation to, probably the same thing that caused people to believe in the old gods and myths, in angels, resurrection, and even UFOs today. It probably starts in the human mind.

Strieber continues this line of thinking in *Transformation*, his 1988 sequel to *Communion*: "It may be, that what happened to Mohammad in his cave and to Christ in Egypt, to Buddha in his youth and to all our great prophets and seers, was an exalted version of the same humble experience that causes a flying saucer to traverse the sky or a visitor to appear in a bedroom."

TLE has changed the course of civilization. LaPlante and others speculate that the mystical religious experiences of many of the great prophets were induced by TLE, because the historical writings describe classic TLE symptoms. Mohammad, Moses, and Saint Paul are the religious prophets most often thought to have had epilepsy. Dostoyevsky thought it was obvious that Mohammad's visions of God had been triggered by epilepsy. "Mohammad assures us in his Koran that he had seen Paradise," Dostoyevsky notes. "He did not lie. He had veritably been in Paradise in an attack of epilepsy, from which he suffered as I do."

When Mohammad first had his visions of God, he felt oppressed, smothered, as if his breath were being squeezed from his chest. Later he heard a voice calling his name, but when he turned to find the source of the voice, no one was there. The local Christians, Jews, and pagan Arabs called him insane. Legend had it that Mohammad had fits as a child. When he was five years old he told his foster parents, "Two men in white raiment came and threw me down and opened up my belly and searched inside for I don't know what." This description is startling similar to the alien abduction experience described by people with TLE such as Whitley Strieber.

Note that the overriding emotion experienced by Mohammad, Moses, and Saint Paul during their religious visions was not one of rapture and joy but rather of fear. In 1300 B.C., when Moses heard the voice of God from a burning bush, Moses hid his face and was frightened. Luke and Paul both agreed that Paul suffered from an unknown "illness" or "bodily weakness," which Paul called his "thorn in the flesh." Famous biblical commentators have attributed this to either migraine headaches or epilepsy. Paul once had malaria, which involves a high fever that can damage the brain. Other psychologists have noted that probable TLEers like Moses, Flaubert, Saint Paul, and Dostoyevsky were also famous for their rages.

The psychologist William James has argued that religious states are not less profound simply because they can be induced by mental anomalies:

> Even more perhaps than other kinds of genius, religious leaders have been subject to abnormal psychical visitations. Invariably they have been creatures of exalted emotional sensitivity liable to obsessions and fixed ideas; and frequently they have fallen into trances, heard voices, seen visions, and presented all sorts of peculiarities which are ordinarily classed as pathological. Often, moreover, these pathological features have helped to give them their religious authority and influence. To plead the organic causation of a religious state of mind in refutation of its claim to possess superior spiritual value, is quite illogical and arbitrary [because] none of our thoughts and feelings, not even our scientific doctrines, not even our *dis*-beliefs, could retain any value as revelations of the truth, for every one of them without exception flows from the state of the possessor's body at the time. Saint Paul certainly once had an epileptoid, if not an epileptic, seizure, but there is not a single one of our states of mind, high or low, healthy or morbid, that has not some organic processes as its condition.

Recently, several nuns with temporal lobe epilepsy have provided further evidence for TLE's being at the root of many mystical religious experiences. Eve LaPlante discusses a former nun who "apprehended" God in the course of her TLE seizures. The nun described the experience: "Suddenly everything comes together in a moment—everything adds up, and you're flooded with a sense of joy, and you're just about to grasp it, and then you lose it and you crawl into an attack. It's easy to see how, in a prescientific age, an epileptic or any temporal lobe fringe experience like that could be thought to be God Himself."

Even Ezekiel in the Old Testament had a TLE-like vision reminiscent of modern UFO and abduction reports:

> And I looked, and behold, a whirlwind came out of the north, a great cloud, and a fire infolding itself, and a brightness was about it, and out of the midst thereof as the color of amber, out of the midst of the fire. . . . Also out of the midst thereof, came the like-

ness of four living creatures. And this was their appearance, they had the likeness of a man. And every one had four faces, and every one had four wings. And their feet were straight feet; and the sole of their feet was like the sole of a calf's foot; and they sparkled like the color of burnished brass. . . . Then the spirit took me up, and I heard behind me a voice of great rushing. . . .

Alien Tests

Intelligence may indeed be a benign influence creating isolated groups of philosopher-kings far apart in the heavens. . . . On the other hand, intelligence may be a cancer of purposeless technological exploitation, sweeping across a galaxy as irresistibly as it has swept across our own planet.

—**Freeman Dyson,** *Scientific American,* **1964**

Maybe the brilliance of the brilliant can be understood only by the nearly brilliant.

—**Anthony Smith,** *The Mind*

If aliens abducted humans to assess our intelligence, what tests would aliens use? Of course, the mere existence of our civilization's artifacts, from skyscrapers to satellites to atomic bombs to Pentium computers, may provide significant information about our level of intelligence. But assume for the moment that aliens want to examine our bodies to learn about us. It is interesting that aliens could not determine which humans are more intelligent than others by measuring the relative sizes of our brains. Human intelligence is not correlated well with brain size. In fact the brain of the great mathematician K. F. Gauss proved to be an embarrassment to those who thought brain size indicated intelligence. Gauss's brain weighed 3.29 pounds (1,492 g), only slightly more than average. However, Gauss's brain was found to be more richly convoluted than average brains. Stephen J. Gould's book *The Mismeasure of Man* gives many examples of large-brained criminals and small-brained men of eminence. The largest female brain ever weighed (3.45 pounds, or 1,565 g) belonged to a woman who killed her husband, and autistic people have heavier brains than average. On the other hand, intelligent people can exist even with ultrathin "potato chip" brains caused by childhood diseases.

Intelligence is not an inevitable result of evolution on any world. Since the beginning of life on Earth, as many as 50 billion species have arisen, and only one of them has acquired technology. If intelligence as such has high survival value, why are so few creatures very intelligent? Mammals are not the most successful of animals. Ninety-five percent of all animal species are invertebrates. Most of the worm species on our planet have not even been discovered yet, and there are a billion billion insects wandering the Earth.

If humankind were destroyed in some great cataclysm, there is very little possibility that our level of intelligence would ever be achieved on Earth again. The historian of science C. Owen Lovejoy regards cognition as a pure accident: "It is evident that the evolution of cognition is neither the result of an evolutionary trend nor an event of even the lowest calculable probability, but rather the result of a series of highly specific evolutionary events whose ultimate cause is traceable to selection of unrelated factors such as locomotion and diet."

If human intelligence is an evolutionary accident, and mathematical, linguistic, artistic, and technological abilities are a very improbable bonus, then there is little reason to expect that life on other worlds will ever develop intelligence as far as we have. Both intelligence and mechanical dexterity appear to be necessary to make radio transmitting devices for communication between the stars. How likely is it that we will find a race having both traits? Very few Earth organisms have much of either trait. As the evolutionary biologist Jared Diamond has suggested in *Natural History*, those that have acquired a little of one (smart dolphins, dexterous spiders) have acquired none of the other, and the only species to acquire a little of both (chimpanzees) has been rather unsuccessful. The most successful creatures on Earth are the dumb and clumsy rats and beetles, which found better routes to their current dominance. All this means that if we *do* receive a message from the stars, it will undermine much of the current thinking on how the evolutionary mechanisms work.

As Dr. W. J. Holland said in his wonderful *The Moth Book*:

When the moon shall have faded out from the sky, and the seas shall be frozen over and the cities have long been dead and crumbled, and all life shall be on the very last verge of extinction; then, on a bit of lichen growing on the bald rocks beside the eternal snows of Panama, shall be seated a tiny insect, preening its antennae in the glow of the worn-out sun, representing the sole survivor of animal life on Earth.

Another interesting point to consider is that when it comes to recognizing internal organs, the average dolphin is more "intelligent" than the average human. With ecolocation, cetaceans can slowly scan the contents of colleagues' intestines, follow the flow of their blood, and maybe access the architecture of their lovers' brains. How would human sexuality be different if we had the ability to see inside one another? Would magazines like *Playboy* have evolved if we could see the "inner beauty" of our lovers?

Another Skeptical Look

Are we dealing instead with a parallel universe, another dimension, where there are human races living, and where we may go at our expense, never to return to the present?

—John Mack

For the first time in our history, we are studying something that is studying us.

—Marilyn Teare, California therapist

In this section, I revisit a few prior topics from an even more skeptical perspective. I'll play the devil's advocate, telling you all the reasons why alien visitations and civilizations seem unlikely. My goal is to encourage heated debate and critical thinking, so for each argument I raise, feel free to come up with counterarguments.

Despite my highly scientific training, the concept of alien abduction continues to haunt me. I'm particularly fascinated by how the physical appearance of reported alien visitors has evolved through time.[3] In the late 1940s, aliens looked like "little green men." In the early 1950s, beautiful humanlike beings appeared to the "contactees." Hairy dwarfs were common in the mid-1950s. The appearance of big-eyed humanoids with large, wraparound eyes begins to appear in 1961, the date of the first widely reported alien abduction—the Betty and Barney Hill case.

If some of you think that alien abductions are initiated by physical beings, you should ask yourself why should these beings appear humanoid? It is possible that alien organisms would evolve with their brains close to their eyes and high up on their body, but it is hard to believe that they would so closely resemble humans as depicted by abductees. Professor Michael Grosso, an artist and philosopher who has taught at several colleges and universities, notes that our images of fetuslike beings mimic the commonly televised pictures of starving children in famines, with their sticklike arms, large heads, and bulging eyes. Could the constant TV exposure explain the persistence and similarity of alien images?

Would aliens be able to breathe our atmosphere? Some scientists believe that civilizations could only evolve on planets with an oxygen content similar to ours. With too little oxygen, a developing culture could not use fires to their advantage, and with too much oxygen, fires would burn out of control. Without fire, would humans have created advanced civilizations?

I find it difficult to believe in literal, physical abductions by extraterrestrials because there is very little definitive physical evidence such as photographs of beings. Why so few witnesses? Why is it that astronomers who watch the sky almost constantly, and photograph it extensively, have never reported a flying saucer? If tiny objects were implanted within abductees' bodies with the frequency reported, why aren't the implants captured on X-rays and CAT scans with a high incidence? How do aliens convey abductees through solid walls and windows?

Despite these questions, some researchers argue that abductions are more than a strange mental phenomenon because many abductees do not exhibit psychiatric disorders, because details are consistent among people who reluctantly tell their stories, and because even very young children claim to have been abducted. On the other hand, some psychologists believe that many "abductees" have low self-esteem and derive a degree of importance simply from becoming the object of interest of an abduction researcher or psychologist.

A small percentage of abductees seem to have an obvious history of mental aberrations, of making exaggerated claims, or of having a fascination with "borderland science" such as psychic phenomena, channeling, and other New Age activities. For example, some abductees claim to see auras and can feel people's "blocked energy vibrations." Some report supernatural experiences including telepathic communication with ghosts, and have had a lifetime of odd "dissociative episodes." One abductee in the literature remembered from her childhood the strange image of a large, gray stone rabbit beside her crib. Another was asked permission by an alien to try on her high-heeled shoes. How are we to distinguish the visions of schizophrenics from more "serious" abduction claims?

Whitley Strieber is perhaps the best-known of the abductees. How are we to assess his encounters when, according to Phillip Klass's *UFO Abductions: A Dangerous Game*, Strieber appears to have had a lifetime of unusual experiences. For example, in April 1977 Strieber held a brief conversation with a voice that had come over his stereo system in his New York apartment. Strieber has also encountered robotlike aliens, aliens with pug noses wearing blue coveralls, an alien with an inept cardboard imitation of a blue double-breasted suit (complete with a white-triangle handkerchief sticking out of the pocket), and tall aliens in tan jumpsuits. One alien had a "ridiculous excuse for a curly black toupee on his head." Strieber once even saw a giant, ugly insect that made him feel mothered and loved. Another time he reported seeing the head of a living woman named Kathie Davis on a shelf inside a UFO.

Strieber has had some other odd experiences. Once when he was eating an ice cream cone, he heard a voice cry out, "Can you stop eating that?" Another time, a UFOnaut told him to stop eating chocolate or he would die. At one point in his life, Strieber purchased a riot gun for protection. When asked why, he replied, "Not sure. I just have the feeling sometimes . . . there are people in the house." At other times he has set up elaborate security systems to protect himself from death threats made by "right-wing southern groups," but then he later admitted that there were never such threats.

Several people during their abductions have been shown (telepathically) hybrid fetuses and children. Some female abductees tell us that their eggs were stolen from them because aliens want to breed with humans. However, the possibility of creating an alien/human hybrid seems far-fetched. If humans can't produce a hybrid creature by fertilizing a chimpanzee's egg with human sperm, how could an alien, with a presumably much different genetic heritage and biochemistry, make use of our DNA?

Finally, what is the chance that alien civilizations exist? My personal view is that civilizations sufficiently advanced to explore outer space or communicate with us are quite rare. For one thing, liquid water is important for life. There are probably few water-containing planets that orbit at the precise distance from a sun so that water does not boil away or perpetually freeze, although icy moons may be common. Primitive life-forms could evolve on cold icy worlds, but technological development would be quite difficult.

Earth's fixed axis of rotation is essential for our own life, because of the stabilizing effect this has on climate. Many planets do not have stable axes. Although we simply do not understand all the factors contributing to climate change and stabilization, the Earth may be quite atypical. The Earth has a very strong magnetic field, which is vital for maintaining the ozone layer protecting life from deadly ultraviolet radiation.

Planetary size can play an important role in fostering the emergence of life. For example, planets much larger than Earth will outgas more material, thereby enhancing the greenhouse effect. Calculations show that if Earth's mass had been 10 percent greater, this outgassing effect would have prevailed and there would have been no orbit in which Earth could have traveled and still retained liquid oceans. On the other hand, if the planet is too small, it will not retain an atmosphere effective in blocking out dangerous solar ultraviolet radiation.

If we did not have our Moon, there would be no tides, and consequently no region of shoreline exposed at low tide and covered at high tide. With a much narrower "intertidal zone," the diversity of life on Earth would be reduced fantastically—and humans would not have evolved. The intertidal zone was a bridge between the land and the sea. Today, certain crustaceans and worms actually thrive in this zone. Without it, the evolutionary transition to land may have never taken place because the water-land edge would be an insurmountable barrier. Thanks to our Moon, life was able to leave the sea about half a billion years ago.

Could intelligent, technologically advanced life develop on a world entirely covered by liquid, or on one that, like Jupiter, consists only of gas? Probably not. For one thing, life seems to need the challenge of land to expand its intelligence. Smart aquatic animals, like the bottlenose dolphin, evolved from a land-dwelling progenitor. Other large aquatic animals, like sharks, are quite dumb. Speaking from a biochemical standpoint, we find that the wet and dry cycles of tidal pools can bind together several precursors of genetic molecules like RNA. These pools could have been the cradle for genetic molecules, and it is likely that these pools also contained lipids that could spontaneously form into primitive cell membranes.

No land, no intelligence—at least not the kind that can develop complex mathematics and nuclear-powered ships. On worlds with only water, there is no fire. Even if dolphins are in some sense as intelligent as humans, they are probably not interested in astronomy, given the fact that they can't even see the stars, except in the occasional moments when they surface to breathe at night. And even if some dolphin philosopher glimpsed the stars while gazing beyond the confines of his water world, he couldn't build a telescope with his flippers—perhaps his ultimate handicap for gaining enlightenment. With just a few bones in the wrong position, the entire universe is forever beyond his grasp. Are there "bones" that have limited humans in their quest for knowledge and a truer understanding of the cosmos? Without land, it is difficult to gaze at the stars and dream of conquering space. We will never receive a communication from an alien aquatic species.

If Earth were just a bit closer to the sun, it would be totally shrouded in clouds. If life had evolved on a world continually shrouded with clouds, would we have any real interest in extraterrestrial intelligence? Imagine aliens on a shrouded world spending their time in contemplation of eternal philosophical truths and mathematics with no interest at all in trying to make contact with other worlds. Perhaps an alien culture is not the least interested in space travel or technological advancement—like the ancient Chinese who knew of the properties of gunpowder but used it for entertainment rather than as a weapon, and who considered all outsiders as crude barbarians with whom they were uninterested in exchanging ideas.

Another reason we may never have alien visitors is that space travel requires large amounts of metal. Imagine an alien world without significant deposits of metal—for example, no iron or copper, nor alloys such as bronze or steel. What type of civilization could evolve on such a world? Would the civilization peak at a Stone Age level using primitive tools, shelters, and weapons? Perhaps they could develop rubber production and develop plastics and polymers that form the basis of technology. With rubber balloons and access to natural helium gas escaping from gas pockets in swamps, it would be possible to achieve flight on such a planet. However, magnetism would be unknown on such a world, and advanced physics theories could not develop that rely on knowledge of magnetism. (Metal is also important because it is a good conductor of electricity, although ceramics and organic polymers could also be conductors.)

Perhaps on a world without metal deposits, horn-extruding creatures could actively shape, form, and alter the composition of the material extruded. The futurist Jim McLean suggests that a society of "neonarwhals" could create complex structures and mechanical devices without ever having discovered fire or metals. In an ocean such as Earth's, rich in metal ions, marine creatures could extrude biological semiconductors and produce wires from metals harvested from the water.

All of these musings suggest that it theoretically possible to develop technology without metal. However, I believe that the properties of metal (strength, flexibility, malleability, and the ability to change to liquid state with easily attainable heat) have made it a basis of our development. Our ability to create advanced spaceships is highly dependent on our metals and alloys. Without such materials, and the ability to manipulate them, aliens would not be able to greet us as we travel beyond Earth.

Divine Quarantine

Even if one uses optimistic predictions for the possibility of the existence of advanced alien civilizations, the chances of an extraterrestrial race making *physical contact* with us is small. The astronomer Gerrit Verschuur of the University of Memphis Department of Physics, believes that if extraterrestrial civilizations are, like ours, short-lived, then there are probably no more than 10 or 20 of them existing at this moment, all a lonely 2,000 light years apart from one another. "We are," says Vershuur, "effectively alone in the Galaxy." This means it is very unlikely that UFOs and aliens are visiting us. In fact, C. S. Lewis, the Anglican lay theologian, proposed that the great distances separating intelligent life in the universe constitute a form of divine quarantine: "The distances prevent the spiritual infection of a fallen species from spreading." If there is a galactic club of aliens, I'd expect to find them toward the center of our Galaxy, where the stars are more tightly packed, and the mean distance between stars is only one light-year, instead of the nine light-years that separate stars in our region of the galaxy.

Our fastest spaceships can travel about one six-thousandth (.006) the speed of light. Not including our Sun, the nearest star to us is Proxima Centauri, which is 4.2 light-years away. Our fastest ships would require 25,000 years to reach this star. Radio messages would take decades to reach our closest neighbors and thousands of years to cross the Galaxy.

All this talk of cosmic loneliness, while humans live in a Galaxy of trillions of stars, evoke memories of the Austrian poet Karl Kraus (1874–1936), who wrote: "One's need for loneliness is not satisfied if one sits at a table alone. There must be empty chairs as well."

I am also reminded of the haunting lines from "Velvet Green," by the eighteenth-century English agriculturist, writer, and inventor Jethro Tull:

> We'll dream as lovers under the stars:
> Of civilizations raging afar.
> And the ragged dawn breaks on your battle scars
> As you walk home cold and alone upon Velvet Green.

 *10*

CONCLUSiON

Once He created the Big Bang . . . He could have envisioned it going in billions of directions as it evolved, including billions of life-forms and billions of kinds of intelligent beings. As a theologian, I would say that the proposed search for extraterrestrial intelligence (SETI) is also a search of knowing and understanding God through his works—especially those works that most reflect Him. Finding others than ourselves would mean knowing Him better.

—Theodore M. Hesburgh, C.S.C., University of Notre Dame

Our present world would certainly be mind-boggling to most people of any previous century. The one constant today is change itself. By facing and anticipating change, we can dilute fear of the unknown and act in ways that are most appropriate for both ourselves and society at large.

—Edward Cornish

Is mankind alone in the universe? Or are there somewhere other intelligent beings looking up into their night sky from very different worlds and asking the same kind of question?

—Carl Sagan and Frank Drake, *Cosmology+1.*

Someday in the not-too-distant future we will find life on other worlds. The fact that life emerged on Earth suggests that it exists in other parts of the cosmos because the elements of which the entire universe is composed are remarkably uniform. If some of the elements have combined in ways that produce life on Earth, it is likely they have combined in similar ways elsewhere. We have every reason to believe that there are other water-rich worlds in the universe with complex organic molecules. This means that there should be many worlds in the Milky Way capable of supporting simple life-forms. Even as you read these words, there must be planets in other galaxies on which life is just emerging or even flourishing. Just as you blink, some new life-form is arising.

Life emerged extremely quickly on Earth. Indeed, life arose just as soon as it possibly could. From this we can conclude that the development of life is easy. Given sufficient time and the proper environment, life will emerge through the inexorable force of the laws of physics and chemistry. Perhaps early stages of life on some planets may not be wildly different from the first one-celled organisms on Earth. Of course, the complex multicelled aliens that may evolve would be very different from us, having followed their own complex and chaotic evolutionary path. Even if our universe turns out to only permit life based on carbon, such a condition places little limitation on what form life may take. For example, on Earth carbon constitutes everything from a beautiful rose to ten-foot-long sulfur-eating worms at the bottom of the ocean.

I believe that the definitive discovery of alien microbes on a water world like Europa would drastically alter our worldview and change our society as profoundly as did the Copernican, Darwinian, and Einsteinian revolutions—particularly if the alien microbe could be shown to have evolved independently of Earth. It would impact religious thought and spur interest in science as never before.

Some religious thinkers seem to believe that attempts to create "life" in a test tube are wrong and against the will of God. Yet we have seen that life is built into the chemistry of the universe, poised to evolve wherever conditions are right. If we discover *advanced* life-forms in the universe, far from demoting humanity to the status of inferior creatures, this discovery would give us reason to believe that we are part of a grander process of cosmic organization and hope.

If intelligent space-faring aliens evolved and we were able to communicate with them, our correspondence could bring us a richer treasure of information than medieval Europe inherited from ancient Greeks like Plato and Aristotle. Just imagine the rewards of learning alien languages, music, art, mythology, philosophy, biology, and even politics. Who would be the aliens' mythical heroes? Are their gods more like the thundering Zeus and Yaweh, or the gentler Jesus and Baha'u'llah?

During our Renaissance, rediscovered ancient texts and new knowledge flooded medieval Europe with the light of intellectual transformation, wonder, creativity, exploration, and experimentation. Another, even more exciting, Renaissance would be fueled by the wealth of alien scientific, technical and sociological information. Interestingly, the spirit of our Renaissance achieved its sharpest formulation in art. Art came to be seen as a branch of knowledge, valuable in its own right and capable of providing both spiritual and scientific images of our position in the universe. Similarly, the Renaissance caused by alien contact would transform art with new ideas, forms, and emotions.

Wouldn't it be a wild world in which to live if alien messages and technology were common—like the computer and telephone? In such a world, it might be possible to manipulate space and time in such a way as to make travel to other worlds easier. As early as Georg Bernhard Riemann (1826–1866), mathematicians have studied the properties of multiply connected spaces in which different regions of space and time are spliced together. Physicists, who once considered this an intellectual exercise for armchair speculation, are now seriously studying advanced branches of mathematics to create practical models of our universe, and to better understand the possibilities of parallel worlds and travel using wormholes and by manipulating time.

Where will we first find extraterrestrial life? Europa, a moon of Jupiter, seems a likely candidate because recent images reveal Europa's frosty surface to be nothing more than an ice cap floating atop an ocean of water. The *Galileo* space probe in 1997 found brown stains on the ice that could conceivably be a mix of hydrogen cyanide and other life-related chemicals. There are also stranger possibilities to consider. For example, on Jupiter's moon Io and on Venus, life might exist in liquid sulfur. Though Io appears dehydrated, planetologists don't rule out the possibility of subsurface water.[1] Neptune's moon Triton, although quite cold, appears heavy with subsurface ice that was once sufficiently warm to flow over the landscape. Dark streaks near the poles suggest that occasional geysering spouted carbon or some other organic material. Saturn's Titan, larger than both Mercury and Pluto, has an atmosphere 60 percent denser than Earth's and forms a photo-chemical haze filled with prebiological chemicals. Although we expect Titan to be quite cold, it is an ideal location to check for ammonia- or hydrocarbon-based life.

Various astrophysicists have speculated wildly on life based entirely on different physical process including plasma life within stars (based on the reciprocal influence of magnetic force patterns and the ordered motion of charged particles); life in solid hydrogen (based on ortho- and para-hydrogen molecules); radiant life (based on ordered patterns of radiation); and life in neutron stars (based on polymer chains storing and transmitting information). It may be difficult to think of these physical processes as being alive and able to organize into complex behaviors, societies, and civilizations. However, when viewed from afar, it is equally hard to imagine that interactions between proteins and nucleic acids could possibly lead to the wondrous panoply and complexity of Earthly life—from majestic blue whales and ancient redwoods to curious, creative humans who study the stars. If you were a silicon alien from another star system, and you had a map of human DNA or a list of our amino acids, could you use it to predict the rise of civilization? Could you have imagined a mossy cavern, a black viper, a retina, a seagull's cry, or the tears of a little girl? Would you have foreseen Beethoven, Einstein, Michelangelo, or Jesus?

Some of you may be wondering why seasoned scientists are interested in inventing and discussing hypothetical alien life-forms. Science works by asking questions and wondering what *might* be. This is the way scientists devise theories and test hypotheses. Scientific science fiction, such as Carl Sagan's *Contact*, teaches the public how science is done and why it should be supported. It teaches us to wonder about the awesome scale of our universe. Science fiction stories about space travel have already inspired humans to travel to the moon. Similarly, stories about aliens inspire us to learn more about life's chemistry and to create new radio listening devices seeking signs of extraterrestrial life. Will space travel stories inspire us to create increasingly potent technology to travel farther in the universe? Will we ever find a way to overcome the Einstein speed limit and make all of space-time our home?

Zen Buddhists have developed questions and statements called *koans* that function as a meditative discipline. Koans ready the mind so that it can entertain new intuitions, perceptions, and ideas. Koans can not be answered in ordinary ways because they are paradoxical; they function as tools for enlightenment because they jar the mind. Similarly, the contemplation of alien life is replete with koans, and that is why this Conclusion poses more questions then it answers. These questions are koans for scientific minds.

I wonder what humanity will discover as it searches for extraterrestrial life during the next century or two. How far will we travel? Around 4 billion years ago, living creatures on Earth were nothing more than biochemical machines capable of self-reproduction. In a mere fraction of this time, humans evolved from creatures like Australopithecines. Today humans have wandered the Moon and have studied ideas ranging from general relativity to quantum cosmology. Who knows what beings we will evolve into? Who knows what intelligent machines we will create that will be our ultimate heirs? These creatures (Figure 10.1) might survive virtually forever, and our ideas, hopes, and dreams may be carried with them. There is a strangeness to the cosmic symphony that may encompass time travel, higher dimensions, quantum superspace, and parallel universes—worlds that resemble our own and perhaps even occupy the same space as our own in some ghostly manner. The astrophysicist Stephen Hawking has even proposed using wormholes to connect our universe with an *infinite* number of parallel universes. The theoretical physicist Edward Witten is working hard on superstring theory, which has already created a sensation in the world of physics because it can explain the nature of both matter and space-time. Our heirs, whatever or whoever they may be, will explore alien worlds to degrees we cannot currently fathom. They will discover that the universe is a symphony of life-forms played in many keys. There are infinite harmonies to be explored.

I believe that SETI, the search for extraterrestrial intelligence, is important and should be funded, even if there is only a slight chance of detecting an extraterrestrial signal. Aside from advancing our knowl-

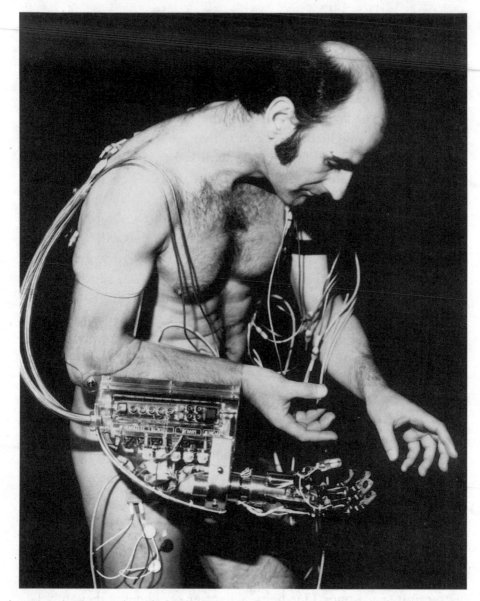

10.1 *Human-machine hybrid. Aliens such as those in the movie* Independence Day *wore a biomechanical exterior that extended their physical abilities and provided protection. Advanced races in the universe may do the
same, and so will we. In the future, our armor will be alive, a second skin
augmented by prostheses and microcomputers. Similarly, we may use computers to augment our brains and extend our imaginations in undreamed-of
ways. (Photo courtesy of Stelarc.)*

edge of computer technology, radio astronomy, communication, chemistry, and biology, SETI is among the wildest adventures in human history. It is our nature to dream, to search, and to wonder about our place in a seemingly lonely cosmos. I agree with Eric Fromm, who wrote in *The Art of Loving*, "The deepest need of man is to overcome his separateness, to leave the prison of his aloneness."

NOtES

Life survives in the chaos of the cosmos by picking order out of the winds. Death is certain, but life becomes possible by following patterns that lead like paths of firmer ground through the swamps of time. Cycles of light and dark, of heat and cold, of magnetism, radioactivity, and gravity all provide vital guides—and life learns to respond to even their most subtle signs. The emergence of a fruitfly is tuned by a spark lasting one thousandth of a second; the breeding of a bristle worm is coordinated on the ocean floor by a glimmer of light reflected from the moon. . . . Nothing happens in isolation. We breathe and bleed, we laugh and cry, we crash and die in time with cosmic cues.

–Lyall Watson, *Supernature*

Preface

1. Basidiomycetes is a large and diverse class of fungi including jelly and shelf fungi; mushrooms, puffballs, and stinkhorns; and the rusts and smuts. They are most often parasites and decomposers. Jew's ear fungus is a brown, gelatinous edible fungus found on dead tree trunks in moist autumn weather.

Introduction

1. A radio telescope consists of a radio receiver and an antenna system used to detect radio-frequency radiation emitted by extraterrestrial sources. Because radio wavelengths are much longer than those of visible light, radio telescopes must be very large in order to attain the resolution of optical telescopes.

2. Alpha Centauri is a triple star, the faintest component of which, Proxima Centauri, is the closest star to the Sun, about 4.3 light-years away.

3. According to J. Talalaj and S. Talalaj's *Strangest Human Sex*, though these ceremonies are no longer official, sometimes the Hiji do them in private

as an expression of their religious beliefs. This discussion of other *cultures* on Earth leads me to wonder about how a planet's *landscape* affects culture. One way to better understand how alien sociology might differ from our own is to imagine hypothetical Earths in which the layout and size of our continents are different.

How would the world be different today, geopolitically speaking, if the ancient land masses had never drifted apart and today's world consisted of a single supercontintent called One World? The diversity of languages would be far less in One World. For example, linguists such as Johanna Nichols from the University of California at Berkeley have done extensive studies reconstructing the spread of prehistoric languages, using comparative linguistics. Languages multiply more rapidly in tropical areas along coastlines and more slowly in the drier interior of continents. The island of New Guinea, for instance, harbors 80 families of languages, the greatest density of languages found anywhere in the world. On the other hand, a much larger region such as Australia only contains about 30 families of languages. If the landmasses of our world had never divided, language diversity would be much less than we have today on our real Earth. Aliens living on a compact dry-land area with few mountains and deserts might develop a single universal language, a development aided by efficient systems of transportation and communication.

Not only language would be affected. If our own supercontinent had never broken up, there would be no totally isolated biomes. Therefore, disparate species such the Australian marsupials, or the Old World and New World primates, would not have evolved.

4. The Arecibo radio telescope is located 10 miles (16 km) south of the town of Arecibo in Puerto Rico. It is the world's largest single-unit radio telescope, built in the early 1960s and employing a 1,000-foot (300 m) spherical reflector.

Chapter 1:
What Aliens Look Like

1. Icthyosaurs are an extinct group of aquatic reptiles that resemble porpoises in appearance and habit.

2. Another example of convergent evolution is aerial flight, developed by the ancestors of birds, insects, and bats, and by the teleost fishes. Photosynthesis was invented by several different bacterialike organisms—violet bacteria, cyanobacteria (the ancestors of green plants), and probably many more forms now extinct.

3. A hyperbeing that lives outside of our three dimensions of time and one dimension of space can effortlessly remove things before our very eyes, giving us the impression that the objects simply disappeared. This is like a 3-D crea-

ture's ability to remove a piece of dirt inside a circle drawn on a page without cutting the circle. The hyperbeing can also see inside any 3-D object or life-form, and if necessary remove anything from inside. The being can look inside our intestines, or remove a tumor from our brain without ever cutting through the skin.

4. J. Travis, "Gene Tells Left from Right," *Science News* 152, no. 4 (1997): 56.

5. The phylum Cnidaria has many body forms, symmetries, colorations, and life histories. Cnidarians are common in tropical waters, and their calcareous skeletons form the reefs in most tropical seas. An alternative name for the phylum is Coelenterata, which refers to these animals' simple organization around a central body cavity (the coelenteron).

6. People like Helen Keller teach us what individuals can accomplish when multiple senses are lost. Keller was an American author and educator who was blind and deaf. However, evolutionarily speaking, the human species could not be intelligent nor space-faring if it never evolved multiple senses.

7. Polarized light is light in which the vibration of the electric or magnetic field is confined to one plane. Ordinary light consists of a mixture of waves vibrating in all directions perpendicular to its line of propagation.

Chapter 2:
Alien Senses

1. "Ultraviolet" refers to that portion of the electromagnetic spectrum extending from the violet (short-wavelength) end of the visible light range to the X-ray region (long wavelengths). Ultraviolet (UV) radiation is undetectable by the human eye, although when it falls on certain materials (such as some minerals), it may cause them to fluoresce, that is, emit electromagnetic radiation of lower energy that we perceive as visible light.

2. Lichens are plants made up of an alga and a fungus growing in symbiotic association on a solid surface such as a rock. Lichens grow slowly. The most common means of reproduction is vegetative; that is, portions of an existing lichen break off and fall away to begin new growth nearby.

Chapter 3:
Life at the Edge

1. The sun formed about the same time as Earth and probably behaved erratically.

2. For an excellent background on eukaryotes in extreme environments, see the Web page http://www.nhm.ac.uk/zoology/extreme.html by Dave Roberts, Department of Zoology, the Natural History Museum, London. For more in-

formation on anaerobes, see Embley et al. (1992), Esteban et al. (1993), Brul and Stumm (1994), and Embley et al. (1995)—listed in the "For Further Reading" section.

3. One of the notable differences among bacteria is their requirement for, and response to, atmospheric oxygen. Whereas almost all eukaryotic organisms require oxygen to thrive, perhaps the majority of bacteria grow well under anaerobic (lacking oxygen) conditions. *Obligate aerobic* bacteria can grow *only* in the presence of oxygen. Bacteria that grow only in the absence of oxygen, such as the methane-producing archaebacteria (methanogens), are *obligate anaerobes*. *Facultative anaerobes* can alternate their metabolic processes depending on the presence of oxygen, using the more efficient process of respiration in the presence of oxygen and the less efficient process of fermentation under anaerobic conditions. Examples of facultative anaerobes include *Escherichia coli* and *Staphylococcus aureus*.

4. See Brock (1978) and Tansey and Brock (1978) for a review of heat-loving eukaryotes.

5. For information on the debate regarding hyperthermophilic eukaryotes, see Forterre et al. (1995), Sprott et al. (1991), Bouthier de la Tour et al. (1991), Stetter et al. (1990), Edmonds et al. (1991), and Ciaramella et al. (1995) in the section "For Further Reading."

6. For the perchlike notothenoids, the dominant group of fish in Antarctica, the antifreeze compounds are produced by a gene that evolved from a gene for a digestive protein. For further information on the genetics, see C. Mlot, "Evolutionary Origins of Fish Antifreeze," *Science News* 151, no. 16 (1997): 237. For more general information on antifreeze compounds and Antarctic fishes, see J. Eastman and A. DeVries, "Antarctic Fishes," in *Life at the Edge,* ed. J. Gould and G. Gould (New York: W. H. Freeman, 1986). Note also that various insects, such as the spruce budworm (a moth larva), contain antifreeze proteins.

7. The boreal forests of North America and Eurasia are broad belts of vegetation that span their respective continents from the Atlantic to Pacific coasts. In North America the boreal forest occupies much of Canada and Alaska.

8. Earthly creatures have two sources of carbon, inorganic compounds and organic compounds. Bacteria that use the inorganic compound carbon dioxide as their carbon source are called autotrophs. Bacteria that require an organic source of carbon, such as sugars, proteins, fats, or amino acids, are called heterotrophs. Many heterotrophs, such as *Escherichia coli* and *Pseudomonas aeruginosa*, synthesize all parts of their cells from simple sugars such as glucose because the organisms are able to use all the necessary biosynthetic pathways. Other heterotrophs have lost some of these biosynthetic pathways and require particular amino acids, nitrogenous bases, or vitamins intact in their environments for growth.

9. For more on psychrophiles, see Smith (1984), Cowling and Smith (1987), and Hughes and Smith (1989).

10. For more on acidophiles, see Brock (1978) and Schleper et al. (1995).

11. For more on alkalophiles, see Curds et al. (1986), Finlay et al. (1987), and Kroll (1990).

12. For more information on halophiles, see Rothschild et al. (1994), Gilmour (1990), Grant (1991), and Brown (1990).

13. For more on barophiles, see Bruun (1977), Marsland (1977), and Douglas (1996).

14. For more information on aridophiles, see Rothschild et al. (1994).

Chapter 4:
Weirder Worlds

1. Generally speaking, black holes are collapsed stars—objects so massive that not even light can escape from their surface. See my book *Black Holes, A Traveler's Guide* (Wiley) for more information on black holes.

2. The notion of proton decay is still controversial, but many physicists believe that these extraordinarily long-lived particles eventually die as a result of baryon nonconservation decay paths. Baryons are heavy subatomic particles such as protons and neutrons.

3. The words "vacuum," "nothing," and "void" usually suggest boring, empty space. However, to modern quantum physicists, the vacuum of space has turned out to be rich with complex and unexpected behaviors where a state of minimum energy permits quantum fluctuations. These fluctuations can lead to the temporary formation of particle-antiparticle pairs that usually destroy themselves soon after their creation because there is no source of energy to give the pair permanent existence. These particles are called "virtual particles," and under certain conditions they may separate, become real pairs with positive mass-energy, and become part of the observable world.

4. Braille symbols are made of one to six raised dots arranged in a six-position matrix. Louis Braille, blinded at the age of three, was encouraged to try to communicate with others as a young boy when he learned of a system called "night writing" consisting of raised dots for use at nighttime battlefield communications. Several years later, in 1824 at the age of 15, he invented the more practical Braille system.

5. It's not clear to me that brown dwarfs would have soil, no more than Jupiter or the Sun does. All elements heavier than helium might fall into the center, leaving a hydrogen-helium atmosphere. Nevertheless, we may still speculate about life on high-gravity worlds in general. Also, even if a brown dwarf does not have soil in the sense of grit, sand, and clay, it could have a soil-like material made of a symbiotic mass of fungi, bacteria, protozoans, insects, plant roots, nematodes, and the like.

6. Neutron stars are extremely dense, compact stars composed primarily of neutrons. Neutron stars are typically about 12 miles (20 km) in diameter but have a mass roughly the same as the Sun's. Thus, their densities are extremely high—about 10×10^{14} times that of water.

Chapter 5
Origin of Alien Life

1. Quoted in Sullivan (1994), p. 81.

2. The chemist John Ross of Stanford University and colleagues at the Max Planck Institute for Biophysical Chemistry in Göttingen, Germany, offer rough blueprints for a hypothetical chemical computer based on reactions. This is described in M. Browne, "Chemists' New Tools: Molecular Wee-saws," *New York Times*, April 28, 1992, p. C1.

3. A carbonaceous chondrite is any stony meteorite containing material associated with life (e.g., hydrocarbons, amino acids, and forms resembling microscopic fossils).

4. Interstellar dust contains organic compounds but does not include the essential building blocks of Earth life such as amino acids, sugars, fatty acids, and the bases of nucleic acids. For more information, see R. Shapiro and G. Feinberg, "Possible Forms of Life in Environments Very Different from the Earth," in *Extraterrestrials: Where are They?* ed. B. Zuckerman and M. Hart (Cambridge University Press: New York, 1995).

5. Silicon-based life seems to require the absence of oxygen because the Si-O (silicon-oxygen) bond is very strong and prevents formation of Si-Si-Si complex chains. Many worlds do not have oxygen, although you wouldn't realize this watching the characters in *Star Trek,* who often beam down to planets with no vegetation, but never need breathing apparatuses.

6. Hydrocarbons are organic compounds (such as acetylene or benzene) containing only carbon and hydrogen and often occurring in petroleum, natural gas, and coal. They serve as fuels and lubricants as well as raw materials for the production of plastics, fibers, rubbers, solvents, explosives, and industrial chemicals.

7. Aromatic molecules contain three pairs of doubly bonded atoms (usually all carbon atoms) bonded together in a hexagon shape. The molecule benzene is the smallest molecule of this type. Because benzene and many larger molecules containing the benzene-ring structure have a strong odor, they have come to be known as aromatic compounds.

8. Rebek's proto-life-form is a two-part molecule called an amino adenosine triacid ester, or AATE. The baby molecule produced in the chemical reaction is held to the parent via hydrogen bonds—the same weak bonds that unite the two long helical strands of a DNA molecule. As a result, the parent and the

baby are easily separated and quickly reproduce. The AATE proto-life-form raises large questions about what the first molecule on Earth was that showed signs of life.

9. See A. Clarke, *2010: Odyssey Two* (New York: Ballantine, 1982), and R. Hoagland, "The Europa Enigma," *Sky and Telescope*, January 1980; Dr. Jastrow was previously with the NASA Institute of Space Studies in New York City.

10. If you wish to learn more about electrorheological fluids, such as those making up the blood of hypothetical Ganymedean wildlife, see R. Ruthen, "Fickle Fluids," *Scientific American*, July 1992, p. 111.

11. J. Maniloff, "Nanobacteria: Size Limits and Evidence," *Science* 275, no. 5320 (1997): 1775.

Chapter 6:
Alien Sex

1. For more information on this, see W. Duellman, "Reproductive Strategies of Frogs," *Scientific American*, July 1992, pp. 80–87.

2. D. Berreby, "Sex and the Single Hermaphrodite," *Discover*, June 1992, pp. 88–93.

3. Personal communication to the author.

4. Personal communication to the author.

Chapter 7:
Communication

1. Searching for radio waves from outer space is like searching for a particular radio station on Earth by turning a dial on a radio receiver. The hard part is knowing what frequency to tune to. Researchers have determined that 1,420 megahertz per second, corresponding to a wavelength of 21 centimeters, is a unique, objective frequency that should be known to every observer in the universe because it most easily penetrates many planetary atmospheres. It also has the least competition from sources elsewhere in the galaxy, apart from the whisperings generated by drifting clouds of hydrogen. And since it reveals the positions and motions of the vast hydrogen clouds that form a major component of the galaxy, it would be scanned by radio telescopes everywhere and be an obvious "meeting place."

2. Technically speaking, a prime number is a positive integer greater than 1, divisible only by itself and 1; 2, 3, 5, 7, 11, 13, 17, 19, 23 are all prime numbers. Here's a BASIC program you can use to find the prime factors for the "Tiberian" (or any other) number:

```
10   REM Find Prime Factors of Tiberian Number:
30   A = 16769021
40   IF ABS(A) <= 1 THEN 210
50   N = INT(ABS(A))
60   REM Find the prime factors and print
70   B=0
80   FOR I = 2 TO N/2
90      IF N/I > INT(N/I) THEN 170
100     B= B+1
110     IF B> 1 THEN 130
120     PRINT "Prime Factors of";N; "are:"
130     PRINT I
140     N=N/I
150     IF N=1 THEN 210
160     I=I-1
170  NEXT I
180  IF N<> INT(A) THEN 120
190  PRINT N; "is a prime number."
210  END
```

3. A transcendental number can be represented as a string of digits that never ends and in which no orderly pattern can be detected. For example, pi (π), the ratio of the circumference of a circle to its diameter, is 3.1415. . . . Human beings know the value of π to over a billion decimal places.

4. For more information on the Voynich manuscript, see my Web site at http://sprott.physics.wisc.edu/pickover/home.htm.

5. Hans Freudenthal's language is explained in his book *Lincos: Design of a Language for Cosmic Intercourse* (Amsterdam: North-Holland Publishing, 1960). The three-letter symbols are derived from Latin roots. For example, "Fem" means "female." "Msc" means "male." Here is a decoding of part of the message: "The existence of the human body begins some time earlier than that of the human itself. The same is true for animals. Mat, mother. Pat, father. Before the individual existence of a human, its body is part of the body of its mother. It has originated from a part of the body of its mother and a part of the body of its father."

6. Alien message 1 marks the position of prime numbers in our number system:

```
0110101000101000101000100001010000010001010  . . .
  | | |    | |    | |    |     | |     |    | |
 23 5 7   11 13  17 19  23    29 31   37   41 43
```

Alien message 2 represents the digits of $e = 2.71828$ . . . (Euler's number, the base of natural logarithms), where the ith term of the sequence is the next i digits of e.

Alien message 3: The solution is 8. To solve this, place a multiply operator (times sign) between two digits: $7 \times 7 = 49$, $4 \times 9 = 36$, $3 \times 6 = 18$, $1 \times 8 = 8$.

Alien message 4 consists of the digits in the decimal portion of π: 3.1415...

Alien message 5 is called a *Morse-Thue sequence*. Whenever you see a 0 you replace it with a 01. Whenever you see a 1 you replace it with a 10. Starting with a single 0, we get the alien sequences. Notice that 0110 is symmetrical, a palindrome, but the next pattern 01101001 is not. But hold on! The very next pattern 0110100110010110 is a palindrome again. Does this property continue to hold for alternate sequences? Your study of this remarkable sequence has only just begun. Although aperiodic, the sequence is anything but random. It has strong short-range and long-range structures. For example, there can never be more than two adjacent terms that are identical. One method for finding patterns in a sequence, the Fourier spectrum, shows pronounced peaks when used to analyze the sequence. The sequence grows very quickly. The following is the sequence for the eighth generation:

0110100110010110100101100110100110010110011010010110100110011001
0110100101100110100101101001100101100110100110010110100100110110
0110100110010110011010010110100110010110011010011001011010010110
0110011010010110100110010110100101100110100110010110011001101001
0110100110010110

When converted to sounds, the rhythm pattern is certainly strange to hear.

7. R. Bracewell, "Communications from Superior Galactic Communities," *Nature*, May 28, 1960, pp. 670–671.

Chapter 8:
Alien Travel

1. An object's mass increases with increasing velocity v, as suggested by the formula

$$m = m_0 / \sqrt{1 - (v/c)^2}$$

where m_0 is the rest mass when velocity v is 0. The variable c is the speed of light, 2.9×10^8 meters per second. (Photons of light have zero mass to begin with and so can move at the speed of light.)

2. Interestingly, in the spring of 1997, the astrophysicist William Purcell of Northwestern University discovered vast plumes of antimatter spewing out from the center of the Galaxy and reaching trillions of miles into space.

3. A. U. Smith, "Studies on Golden Hamsters During Cooling to and Rewarming from Body Temperatures Below 0 Degrees Centigrade," *Proceedings of the Royal Society, Biology* (London), *Series B*. 147 (1957): 517.

4. I. Suda and A. C. Kito, "Histological Cryoprotection of Rat and Rabbit Brains," *Cryoletters* 5 (1966): 33.

Chapter 9:
Alien Abduction

1. *Jamais vu* is the feeling of never having been in what should be a familiar place—the opposite of *déjà vu*. Whitley Strieber remarks in *Communion*, "The corridor into our world could in a very sense be through our own minds. Maybe really skilled observation and genuine insight will cause the visitors to come bursting to the surface shaking like coelacanths in a net. Something is here, be it a message from the stars or from the booming labyrinth of the mind ... or from both."

2. Formication is the feeling of bugs crawling under the skin.

3. For a fascinating overview of the look of aliens through time, see J. Nickell, "Extraterrestrial Iconography, *Skeptical Inquirer* 21, no. 5 (Sept.–Oct.): 18–19.

Chapter 10:
Conclusion

1. Erupting volcanoes rise from Io's surface, where the gravitational influence of three nearby moons is enough to distort the shape of the world itself, causing it to pulse like a heart.

FOR FURtHER READING

The shift demanded by the UFO phenomenon that is so difficult is the one that forces us to consider that there might exist simultaneous other realities; further, that it is during, or within, some sort of overlapping of these realities that alien abductions occur.

—C. D. B. Bryan

Aldrin, B., and J. Barnes. *Encounter with Tiber*. New York: Warner Books, 1996.

Bada, Jeffrey L. "Extraterrestrial Handedness." *Science* 275, no. 14 (1997): 942–943.

Barlowe, W., I. Summers, and B. Meacham. *Barlowe's Guide to Extraterrestrials*. New York: Workman, 1979.

Baross, J. A., and J. F. Holden. "Overview of Hyperthermophiles and Their Heat-Shock Proteins." In *Advances in Protein Chemistry*, vol. 48, ed. M. W. Adams. New York: Academic Press, 1996, pp. 1–35.

Blackmore, S. "Alien Abduction." *New Scientist*, Nov. 19, 1994, pp. 29–31.

Bouthier de la Tour, C., et al. "Reverse Gyrase Is Present in Thermophilic Eubacteria." *Journal of Bacteriology* 173 (1991): 3921–3923.

Brock, T. D. *Thermophilic Microorganisms and Life at High Temperatures*. New York: Springer Verlag, 1978.

Brown, A. D. *Microbial Water Stress Physiology: Principles and Perspectives*. New York: John Wiley, 1990.

Brul, S., and C. K. Stumm. "Symbionts and Organelles in Anaerobic Protozoa and Fungi." *Tree* 9 (1994): 319–324.

Bruun, A. F. "Animals of the Abyss." In *Conditions for Life: Readings from Scientific American*, ed. A. Gibor. San Francisco: W. H. Freeman, 1977, pp. 208–215.

Bryan, C. D. B. *Close Encounters of the Fourth Kind: Alien Abduction, UFOS, and the Conference at M.I.T.* New York: Knopf, 1995.

Casti, J. *Pardigms Lost*. New York: Morrow, 1990.

Chown, M. "The Alien Spotters." *New Scientist*, April 19, 1997, pp. 29–31.

Ciaramella, M., et al. "Molecular Biology of Extremophiles." *World Journal of Microbiology* 11 (1995): 71–84.

Cohen, J. "How to Design an Alien." *New Scientist*, Dec. 21, 1991, pp. 18–21.

Cowen, R. "From Here to Eternity: Tracking the Future of the Cosmos." *Science News* 151, no. 14 (1997): 209–209.

Cowling, A. J., and H. G. Smith. "Protozoa in the Microbial Communities of Maritime Antarctic Fellfields." *Colloque sur les Ecosystèmes Terrestres Subantarctiques,* no. 58. Paimpont: Comité National Français des Recherches Antarctiques, 1987, pp. 205–213.

Cronin, John R., and Sandra Pizarello. "Enantiometric Excesses in Meteoritic Amino Acids." *Science* 275 (Feb. 1997): 951–955.

Curds, C. R., S. S. Bamforth, and B. J. Finlay. "Report on the Freshwater Workshop in Kisumu, Kenya (June 30 – July 5, 1985)." *Insect Science Applications* 7 (1986): 447–449.

Davies, P. *Are We Alone?* New York: Basic Books, 1995.

Douglas, A. E. "Microorganisms in Symbiosis: Adaptation and Specialization." In *Evolution of Microbial Life*, no. 54, ed. D. McL. Roberts et al. Society for General Microbiology Symposium, 1996, pp. 225–242.

Drake, Frank. "Summary of the Conference." In *Astronomical and Biochemical Origins and the Search for Life in the Universe: Proceedings of the 5th International Conference on Bioastronomy* (IAU Colloquium no. 161, Capri, July 1–5, 1996), ed. Cristiano Batalli Cosmovici, Stuart Bowyer, and Dan Werthimer. Editrice Compositori, 1997, pp. 789–794.

Drake, Frank, and Dava Sobel. *Is Anyone Out There?* New York: Delta, 1992.

Dyson, F. "Time Without End: Physics and Biology in an Open Universe." *Reviews of Modern Physics* 51, no. 3 (1979): 447–460.

Edmonds, C. G., et al. "Posttranscriptional Modification of Trna in Thermophilic Archaea (Archaebacteria)." *Journal of Bacteriology* 173 (1991): 3138–3148.

Embley, T. M., et al. "The Use of Rrna Sequences and Fluorescent Probes to Investigate the Phylogenetic Positions of the Anaerobic Ciliate *Metopus palaeformis* and Its Archaeabacterial Endosymbiont." *Journal of General Microbiology* 138 (1992): 1479–1487.

———. (1995) "Multiple Origins of Anaerobic Ciliates with Hydrogenosomes Within the Radiation of Aerobic Ciliates." *Proceedings of the Royal Society* (London) 262 (1995): 87–93.Emery, C. "John Mack: Off the Hook at Harvard, but with Something Akin to a Warning." *Skeptical Inquirer* 19, no. 6 (Nov.–Dec. 1995): 4–5.

———. "Alien Autopsy: Show-and-Tell." *Skeptical Inquirer* 19, no. 6 (Nov.–Dec. 1995): 15–16.

Esteban, G., et al. "New Species Double the Diversity of Anaerobic Ciliates in a Spanish Lake." *FEMS Microbiology Letters* 109 (1993): 93–100.

Finlay, B. J., et al. "Ciliated Protozoa and Other Microorganisms from Two African Soda Lakes (Lake Nakuru and Lake Simbi, Kenya)." *Archiven Protistenkunde* 133 (1987): 81–91.

Forterre, P. "Thermoreduction, a Hypothesis for the Origin of Prokaryotes." *Science de la vie/Life Sciences* (C.R. Academy of Science, Paris) 318 (1995): 415–422.

Forterre, P., et al. "Speculations on the Origin of Life and Thermophily: Review of Available Information on Reverse Gyrase Suggests That Hyperthermophilic Prokaryotes Are Not So Primitive." *Origin of Life and Evolution of the Biosphere* 25 (1995): 235–49.

Forward, R. "When You Live upon a Star. . . " *New Scientist,* Dec. 24, 1987, pp. 36–38.

Frazier, K. (1995) "UFOs Real? Government Covering Up? Survey Says 50 Percent Think So." *Skeptical Inquirer 19, no.* 6 (Nov.–Dec. 1995): 3–4.

Freudenthal, H. *Lincos: Design of a Language for Cosmic Intercourse.* Amsterdam: North-Holland Publishing, 1960.

Gamow, G. *One, Two, Three . . . Infinity.* New York: Dover, 1947.

Gilmour, D. "Halotolerant and Halophilic Microorganisms." In *Microbiology of Extreme Environments,* ed. C. Edwards. Milton Keynes, U.K.: Open University Press, 1990, pp. 147–177.

Grant, W. D. "General View of Halophiles." In *Superbugs: Microorganisms in Extreme Environments,* ed. K. Horikoshi and W. D. Grant. Tokyo: Japan Scientific Societies Press, 1991, pp. 15–37.

Haeckel, E. *Art Forms in Nature.* New York: Dover, 1974.

Harter, J. *Animals.* New York: Dover, 1979.

Heidmann, J. *Extraterrestrial Intelligence.* New York: Cambridge University Press, 1995.

Hopkins, B. *Intruders.* New York: Ballantine, 1987.

Hoyle, F. *The Black Cloud.* New York: Harper, 1957.

Huber, R. *Treasurer of Fantastic and Mythological Creatures.* New York: Dover, 1980.

Hughes, J., and H. G. Smith. "Temperature Relations of *Heteromita globosa* Stein in Signy Island Fellfields." *Antarctic Special Topics* (1989):117–122.

Jonas, D., and D. Jonas. *Other Senses, Other Worlds.* New York: Stein and Day, 1976.

Kerr, R. Life Goes to Extremes in the Deep Earth—and Elsewhere? *Science* 276, no. 313 (1997): 703.

Klass, P. *UFO Abductions: A Dangerous Game.* Buffalo, N.Y.: Prometheus Books, 1994.

Kroll, R. G. "Alkalophiles." In *Microbiology of Extreme Environments,* ed. C. Edwards. Milton Keynes, U.K.: Open University Press, 1990, pp. 55–92.

———. "The GAO Roswell Report and Congressman Schiff." *Skeptical Inquirer* 19, no. 6 (Nov.–Dec. 1995): 20–22.

LaPlante, E. *Seized.* New York: HarperCollins, 1993.

Mack, J. *Abduction.* Rev. ed. New York: Ballantine, 1995.

Marsland, D. "Cells at High Pressure." In *Conditions for Life: Readings from "Scientific American,"* ed. A. Gibor. San Francisco: W. H. Freeman, 1977, pp. 200–207.

Maruyama, M., and A. Harkins. *Cultures Beyond Earth.* New York: Vintage Books, 1975.

Monastersky, R. "Deep Dwellers: Microbes Thrive Far Below Ground." *Science News* 151, no. 13 (1997): 192–193.

Morrell, V. "Tracing the Mother of All Cells." *Science* 276, no. 5313 (1997): 70.

————. "Microbiology's Scarred Revolution." *Science* 276, no. 5313 (1997): 699.

Nichols, P. *The Science in Science Fiction.* New York: Knopf, 1983.

Nickell, J. "Alien Autopsy Hoax." *Skeptical Inquirer* 19, no. 6 (Nov.–Dec. 1995)): 17–19.

Pennisi, E. "In Industry, Extremophiles Begin to Make Their Mark." *Science* 276, no. 5313 (1997): 705.

Pickover, C. *Chaos in Wonderland.* New York: St. Martin's Press, 1995.

————. *Mazes for the Mind.* New York: St. Martin's Press, 1993.

————. *Black Holes: A Traveler's Guide.* New York: John Wiley, 1996.

————. *The Alien IQ Test.* New York: Basic Books, 1997.

————. *The Loom of God.* New York: Plenum, 1997.

————. *Strange Brains and Genius.* New York: Plenum, 1998.

Rothschild, L. J., et al. "Metabolic Activity of Microorganisms in Evaporites." *Journal of Phycology* 30 (1994): 431–438.

Schleper, C., et al. "Life at Extremely Low pH." *Nature* (London) 375 (1995): 741–742.

Simons, G. *Simons' Book of World Sexual Records.* New York: Bell Publishing, 1975.

Smith, H. G. "Protozoa of Signy Island Fellfields." *British Antarctic Survey Bulletin,* no. 64 (1984): 55–61.

Sprott, G. D., et al. "Proportions of Diether Macrocyclic Diether and Tetraether Lipids in *Methanococcus janaschii* Grown at Different Temperatures." *Journal of Bacteriology* 173 (1991): 3907–3910.

Stetter, K. O., et al. "Hyperthermophilic Microorganisms." *FEMS Microbiology Review* 75 (1990): 117–124.

Strieber, W. *Communion.* New York: Avon, 1987.

Sullivan, W. *We Are Not Alone.* Rev. ed. New York: Plume, 1994.

Tansey, M. R., and T. D. Brock. "Microbial Life At High Temperatures: Ecological Aspects." In *Microbial Life in Extreme Environments,* ed. D. J. Kushner. London: Academic Press, 1978, pp. 159–194.

Wu, C. "Sometimes a Bigger Brain Isn't Better." *Science News* 148, no. 8(1995): 116.

Zuckerman, B., and M. Hart. *Extraterrestrials: Where Are They?* New York: Cambridge University Press, 1995.

iNDEX

ABOUT
THE AUtHOR

Clifford A. Pickover received his Ph.D. from Yale University's Department of Molecular Biophysics and Biochemistry. He graduated first in his class from Franklin and Marshall College, after completing the four-year undergraduate program in three years. He is the author of the popular books *Time, A Traveler's Guide* (Oxford University Press, 1998), *Strange Brains and Genius* (Plenum, 1998), *The Alien IQ Test* (Basic Books, 1997), *The Loom of God* (Plenum, 1997), *Black Holes–A Traveler's Guide* (John Wiley, 1996), and *Keys to Infinity* (John Wiley, 1995). He is also the author of numerous other highly acclaimed books, including *Chaos in Wonderland: Visual Adventures in a Fractal World* (1994), *Mazes for the Mind: Computers and the Unexpected* (1992), *Computers and the Imagination* (1991) and *Computers, Pattern, Chaos, and Beauty* (1990), all published by St. Martin's Press. He has also written over two hundred articles on topics in science, art, and mathematics. With Piers Anthony, he is the coauthor of the highly acclaimed science fiction novel *Spider Legs*.

Dr. Pickover is currently an associate editor for the scientific journals *Computers and Graphics*, *Computers in Physics*, and *Theta Mathematics Journal*; is an editorial board member for *Speculations in Science and Technology*, *Idealistic Studies*, *Leonardo*, and *YLEM*; and has been a guest editor for several scientific journals. He was the editor of *Chaos and Fractals: A Computer Graphical Journey* (Elsevier, 1998); *The Pattern Book: Fractals, Art, and Nature* (World Scientific, 1995); *Visions of the Future: Art, Technology, and Computing in the Next Century* (St. Martin's Press, 1993); *Future Health* (St. Martin's Press, 1995); *Fractal Horizons* (St. Martin's Press, 1996); and *Visualizing Biological Information* (World Scientific, 1995); and coedited *Spiral Symmetry* (World Scientific, 1992) and *Frontiers in Scientific Visualization* (John Wiley, 1994). Dr. Pickover's primary interest is in finding new ways continually to expand creativity by melding art, science, mathematics, and other seemingly disparate areas of human endeavor.

The *Los Angeles Times* recently proclaimed, "Pickover has published nearly a book a year in which he stretches the limits of computers, art and thought." He received first prize in the Institute of Physics' "Beauty of Physics Photographic Competition". His computer graphics have been featured on the covers of many popular magazines, and his research has recently received considerable attention by the press and broadcast media—including CNN's *Science and Technology Week,* the Discovery Channel's program "Understanding Beauty," *Science News,* the *Washington Post, Wired,* and *The Christian Science Monitor*—and also in international exhibitions and museums. *OMNI* magazine recently described him as "Van Leeuwenhoek's twentieth century equivalent." *Scientific American* has featured his graphic work several times, calling it "strange and beautiful, stunningly realistic." Dr. Pickover has received U.S. Patent 5,095,302 for a 3-D computer mouse and 5,564,004 for strange computer icons.

Dr. Pickover is currently a research staff member at IBM's T. J. Watson Research Center, where he has received 15 invention achievement awards, three research division awards, and five external honor awards. Dr. Pickover is also a novelist and is the lead columnist for the brain-boggler column in *Discover* magazine.

Dr. Pickover's hobbies include the practice of Chang-Shih tai chi chuan (a form of martial arts) and shaolin kung fu, raising golden and green severums (large tropical fish found in the central Amazon basin), and piano playing (mostly jazz). He is also a member of the SETI League, a worldwide group of radio astronomers and signal-processing enthusiasts who systematically and scientifically search the heavens to detect evidence of intelligent extraterrestrial life. He can be reached at P.O. Box 549, Millwood, New York 10546–0549, U.S.A. Visit his Web site, which has received over 200,000 visits: *http://sprott.physics.wisc.edu/pickover/home.htm.*